LIFE OF EARTH

LIFE OF EARTH: PORTRAIT OF A BEAUTIFUL, MIDDLE-AGED, STRESSED-OUT WORLD

Stanley A. Rice

59 John Glenn Drive
Amherst, New York 14228-2119

Published 2011 by Prometheus Books

Inquiries should be addressed to
Prometheus Books
59 John Glenn Drive
Amherst, New York 14228–2119
VOICE: 716–691–0133
FAX: 716–691–0137
WWW.PROMETHEUSBOOKS.COM

15 14 13 12 11   5 4 3 2 1

Library of Congress Cataloging-in-Publication Data

Rice, Stanley A.
Life of earth : portrait of a beautiful, middle-aged, stressed-out world / Stanley A. Rice
p. cm.
Includes bibliographical references and index.
ISBN 978–1–61614–225–4 (cloth : alk. paper)
1. Evolution (Biology) 2. Ethics, Evolutionary. I. Title.

QH366.2R5236 2010
570—dc22

2010029639

Printed in the United States of America on acid-free paper

I dedicate this book to Lynn Margulis, Distinguished Professor of Geosciences at the University of Massachusetts at Amherst. Lynn pioneered some of the most important insights in modern evolutionary science, particularly regarding the role of symbiosis in the origin of evolutionary innovations. Hearing a scientific presentation by Lynn, when I was an undergraduate at the University of California, Santa Barbara, was one of the mind-expanding events that led me into a scientific career.

CONTENTS

LIST OF ILLUSTRATIONS

INTRODUCTION

A Pile of Rocks in the Middle of Kansas

In the middle of Kansas wheat fields, a pile of rocks has been cemented together into a pedestal. Nothing in particular distinguishes this place, except that it happens to be the geographic center of the United States (figure I-1).

Figure I-1. This pedestal of rocks near Smith Center, Kansas, marks the geographical center of the United States and the metaphorical midpoint of the life of Earth, the solar system, and the universe. Photograph by the author.

Leave aside for the moment the question of how a country with fractally irregular coastlines can have a geographic center. This designation gives the impression that if the United States were a pancake, it could be balanced on a fork right underneath this pedestal. (Some scientists with too much time on their hands actually calculated the surface irregularities of a pancake and determined that if the pancake were the size of Kansas, it would have more topographical relief; therefore, they concluded, Kansas is flatter than a pancake.)[1] But do you take into account the additional weight of the Rocky Mountains, Sierra Nevada, and Cascades? Do you include the parts of the continent (continental shelves) that are under shallow ocean water? Do you measure the shorelines at high or at low tide? The only thing for certain is that the additional weight of buildings and people is negligible compared to the overwhelming immensity of the landscape. Whatever the case, somebody somewhere decided that a spot outside of Smith Center, Kansas, was the center of the coterminous United States and built a pedestal there, with a plaque on it. In this book, I will use this monument as an anchor—it will represent the present on a universal timescale.

THE BIG BANG

Most people have heard of the big bang, the beginning of the universe.[2] About 13.7 billion years ago, the entire universe may have been contained in an infinitely small volume at infinite density. This point contained what would later be all of the energy, matter, space, and time in the universe. Astrophysicists call this point, with astonishing understatement, a singularity. This point then exploded, spewing out dense, hot plasma, which was not yet energy or matter. "Big bang" is an almost childish name for this event. Indeed, it was a term, possibly derogatory, invented in 1949 by astronomer Fred Hoyle, who did not believe that it had occurred.

The discoveries that led to the big bang theory took astronomers by surprise. Most scientists, like most people, assumed that the universe must have been around forever. They thought of the universe the way the eighteenth-century geologist James Hutton thought of Earth: no vestige of a beginning, no prospect of an end. Religious people believed that God created the cosmos and would then destroy it at times attributable only to God's mysterious will. To believers, the universe was unchanging because there simply had not been enough time since the creation for much to have happened to it. Although most religious people today simply identify the big bang with the beginning of God's creation, fundamentalists consider the big bang to be a threat to their faith. In my rural Oklahoma town, a church put this message on its marquee: "Big Bang Theory? You've got to be kidding!—God" (figure I-2). The message was apparently so well received that, a few months after they took the message down, they put it back up.

Figure I-2. A church in Durant, Oklahoma, considers the big bang to be a threat to their religious beliefs. Photograph by the author.

Vestiges of Creation

But there are, in fact, vestiges of the beginning of the universe—leftover bits of evidence of the big bang.

Astronomer Edwin Hubble (with a lot of help from his assistant Henrietta Swan Leavitt) figured out a way to estimate the distances from Earth to galaxies outside our own Milky Way galaxy.[3] "Variable stars" are stars that change their light intensity on a regular schedule, in some cases by expanding and contracting every few days or weeks. A certain type of variable star, the Cepheid variables, appeared to all have the same pattern of light intensity throughout our galaxy. When Hubble found Cepheid variable stars in photographs he had made of distant galaxies, he measured their light intensity and calculated their distance from Earth. The distances, it turns out, are staggeringly large, measurable not just in light years but in millions of light years. This was one of the first insights that the universe is unimaginably huge.

Hubble also ran into an unexpected finding. He realized that in the light from these distant galaxies, the absorption lines caused by certain elements such as hydrogen were not where they were supposed to be on the rainbow spectrum. They were shifted toward the red end of the spectrum. This meant that the waves of light from these galaxies were being stretched out. A blue shift would mean the opposite—that the light from the galaxies was being squished. This same phenomenon (the Doppler Effect) happens with sound in the atmosphere: the sound of a car horn moving away from you becomes lower, and the sound of the horn moving toward you becomes higher, because of the stretching and squishing of sound waves. This could only mean that the galaxies are flying away from us, and from one another. The universe is expanding.

And there was more. The more distant galaxies had a greater red shift. That is, the more distant galaxies are moving more rapidly away from us and from one another. This had to mean that all the galaxies originated at one point sometime in the distant past. A simple calculation provided an estimate of when the universe began, the age of the cosmos. The currently accepted age of the universe is about 13.7 billion years.

Absolute zero is the temperature that matter would have if all of its atoms were perfectly still. Scientists have recently produced refrigeration devices that chill atoms so close to absolute zero that images of them can be produced by special microscopes. It is not uncommon to see images of atoms in science magazines. A photograph is an image produced by light waves, but atoms are smaller than the wavelengths of light. Images of atoms are produced instead by magnetic fields. One might expect the temperature of outer space to be absolute zero, but astronomers such as George Gamow calculated that the big bang would have left residual photons (the weightless particles of which light consists) that, after thirteen billion years, would have a large wavelength corresponding to a temperature of about three degrees above absolute zero. Two physicists, Arno Penzias and Robert Wilson, were able to measure this background radiation that permeated the entire sky.[4] Here was evidence that the universe had been born in a gigantic explosion.

Birth and Death of Stars

Galaxies condensed from the primordial plasma just a few million years after the big bang. In fact, the oldest light that the Hubble Space Telescope can detect—light that was created 13.2 billion years ago—seems to show galaxies with stars.

Stars form when gravity causes gas molecules to condense so much that nuclear fusion begins.[5] Hydrogen atoms are crushed so closely together that they become helium and release huge amounts of energy. This energy keeps the fusion reaction going. Large stars burn quickly. The star Eta Carinae, which is about 150 times as massive as the sun, will probably have a total life span of about three million years. Such massive stars use up much of their hydrogen fuel quickly and explode as supernovae. The enormous temperatures and pressures of supernovae allow fusion reactions to create not just helium but many heavier elements, even those heavier than uranium. Our sun, being much smaller, will probably last a total of ten billion years, over three thousand times as long as Eta Carinae. Since the first galaxies formed, many stars have burned out, and many new ones have formed.

The End

And there is, in addition to vestiges of a beginning, also the prospect of an end. The beginning of the universe occurred at a very specific point in time. The end of the universe, in contrast, cannot be precisely calculated. Astronomers once speculated that the expansion of the universe would slow down, due to the pull of gravity among the galaxies, and that this gravitational pull would bring them all back together into a "Big Crunch." However, a closer calculation of the speeds with which the galaxies are moving apart shows that over time the universe has expanded faster and faster. Something—perhaps the dark matter and dark energy that we cannot detect—is pushing the universe apart instead of pulling it back together. Therefore, the universe may just keep expanding, until everything reaches absolute zero (the Big Freeze), or until all matter becomes a uniform thin cloud (the Heat Death). Or perhaps the very atoms themselves will disintegrate (the Big Rip). The universe, therefore, may not die at a precise time but will slowly disintegrate, and do so nearly forever.[6]

As the galaxies keep moving apart during the next thirteen and a half billion years, they may become invisible from ours. Also by that time, many stars will have burned out. The sky will have a lot fewer stars in it, though still more than a person could count on a summer night. For the purposes of this book, I will use thirteen and a half billion years from now as a rough estimate of the end of a universe that will keep decaying a long time afterward. By this calculation, the universe is about halfway through its life.

It will not take that long, however, for Earth and the solar system to end. The sun ignited about five billion years ago. After another five billion years, enough hydrogen will have been used up in fusion reactions that the sun's core will contract and its outer plasma will expand. The expanding outer plasma will form a red giant star that, while cooler than most stars, will be hot enough to strip away all air and water on at least the inner planets such as Venus, Earth, and Mars, thus destroying all life that may still remain on Earth. Even before that time, Earth may have lost most of its water molecules. Solar radiation is already stripping some of our atmosphere away.[7] The core of the sun, which will no longer have thermonuclear reactions, will become a white dwarf star that may take trillions of years to cool off, first into a black dwarf star, and finally into an invisible, cold mass of cinders.

A MIDDLE-AGED PLANET

Back to the pile of rocks in Kansas. If you took a road trip from Norfolk, Virginia, across the United States as directly as major roads would permit, except for a side trip on little roads to Smith Center, you would reach San Francisco after a journey of 3,138 miles. Let us assume that this journey is the equivalent of a twenty-seven-billion-year life span of the universe, and that the universe, sun, and Earth are halfway through their lives. At this scale, each mile would represent a little over four million years.

If the big bang happened in Norfolk, you would not even have gotten to Richmond before the galaxies and stars began to form. But you would drive all the way to central Illinois, perhaps around Effingham—a drive that would take you the equivalent of nine billion years—before you would have seen the dusty beginning of our present solar system. What was happening during the intervening nine billion years? The answer is that our solar system is a second-generation product of an earlier star. During those nine billion years, a gigantic star was born, burned out, and exploded in a supernova that produced the cloud of dust and gas that formed our solar system. The supernova explosion produced enough heavy elements in its death cloud that our solar system could have rocky planets revolving around the newborn sun.

By the time you got to Kansas City, the solar system would have changed from a swirling dust disc with a dim thermonuclear reaction in the center to something that looked like the modern planets revolving around the sun. The dust would have coagulated into planets, and the largest planets, especially Jupiter, would have swept up the remaining asteroids, except for one belt of them beyond Mars. The Earth would have had oceans.

Almost right away, these oceans contained living cells. But you would have driven all the way through Salina, Kansas, and up to US Highway 36, near Belleville, the equivalent of three billion years, before you would have noticed complex life-forms. The entire history of multicellular life would take place on the sixty miles of rural highway between there and Smith Center. Why did it take almost three billion years for organisms with structural complexity to evolve from the first simple cells? During those three billion years, complex chemical reactions evolved within the cells, but it was not until these cells began to merge together that structural complexity arose.

After fifteen miles, you would have seen the first plants and animals on the land surface of Earth. After fifteen more miles, you would have seen Earth nearly die during the great Permian extinction. Only a few miles later would you have seen the world of the dinosaurs. The great Cretaceous extinction, when the asteroid fell out of the sky, would occur by the time you were about fifteen miles away from Smith Center, just outside of Mankato. You would be less than a mile away from the rock pedestal before you would have seen the first human ancestors, still the size of chimpanzees, walking upright. The modern human species came into existence only about one hundred thousand years ago, which is 0.1 million years. You would now be heading down the little country road, within 125 feet of the pedestal.

You are now standing five feet away from the pedestal. You are staring at the entirety of human civilization, from the time dirt-brick towers rose above the plains of Akkad. Take a couple of steps. The Roman Empire. Hold your thumb about three inches away from the pedestal. This is when the Declaration of Independence was signed. The width of your thumb—this is the brief shining moment of your life, compared to the duration of the universe.

Now look west. Somewhere near Grand Junction, in a little valley in the Rockies, the sun will expand into a redness that will cleanse away all life on this planet. And in the far distance, San Francisco sits on the edge, beyond which the universe will have expanded so much that no other galaxies may be visible and many stars will have burned out.

Your life is almost insignificant in comparison to that of the living Earth. And yet the human species, during such a tiny fraction of Earth's life span, has achieved a largely complete understanding of the universe, life, and its history, and has also wrought great devastation on Planet Earth. This book is about the processes of evolution, such as altruism and sexual selection. And humans represent the most extreme manifestation of some of these processes. Whether we deserve it or not—and we almost certainly don't—we are tiny stewards of the immense whole of Earth. At least we should know what we are stewards of.

A BEAUTIFUL AND STRESSED-OUT PLANET

Planet Earth is the home of a living network of cells and organisms. Not just the organisms but the network itself is alive. This network has altered the atmosphere, water, and crust of Earth into a form that is recognizably different from what it would have been had life not come into existence. In particular, photosynthetic organisms have used the energy of sunlight to make their food, and as a result have produced an atmosphere that contains a lot of oxygen. Oxygen is very reactive. All of the oxygen would have quickly reacted with other molecules and been depleted from the atmosphere were it not continually renewed by photosynthesis. The network of life, and the ways in which life has transformed the Earth: this is what we mean when we say that Earth is a living planet.

Earth is beautiful in its luxurious middle age not just because of its nonliving components such as mountains and waterfalls, but even more so because of its network of life. It is that living matrix, especially trees and other plants, that puts most of the beauty and color into the world and makes one place different from another. Congo would look like Canada, and Brazil would look like Baluchistan, were it not for plants. Lakes would still be blue, and the Painted Desert of Arizona would still have yellow and red stains, but it is plants that make Earth a green planet and cover it, at various places and times, with carpets or mists of floral hues.

Delusions of Grandeur

One of Earth's species, humans, imagines itself to be the center and purpose of the entire universe. In traditional Christianity, humans were the crown of God's creation. One reason the Christian West believed this was that our religious doctrines told us that it was so. Another reason is that when we looked up in the sky, it appeared to us that the sun, moon, and stars were whirling around us. There were lots of stars visible to the unaided eye, but not more than our religions had told us to expect from a heavenly multitude. The Earth as the center of creation was not just a Christian view but was the accepted scientific view at the time of the Greek mathematician Ptolemy, who formalized the concept of geocentrism.

But astronomers found that the heavenly bodies did not move quite like what a geocentrist model would lead us to expect. In particular, the planets (Greek for "wanderers") were moving in confusing ways. Astronomers attributed complex motions to the planets in order to make them fit into a Ptolemaic model. But it became increasingly clear that the only sensible explanation was that the planets, including Earth, revolved around the sun. The Polish astronomer Mikołaj Kopernik (Nicolaus Copernicus) wrote about this heliocentric model, and Italian astronomer Galileo Galilei defended it.[8]

Heliocentrism was a threat to religious views, and therefore to the political power of church leaders, precisely because it demoted humans from the center of the cosmos. If Earth is merely the third planet from the sun, how could we humans be the center of God's attention?

Since the time of Galileo, most people have gotten used to Earth being one of several planets and to the sun being just one of billions of stars in one of billions of galaxies. Very few people insist on geocentrism anymore. There is a website that claims not only that the universe revolves around Earth but that God has an invisible giant electromagnet that suspends Earth to keep it from falling to the bottom of the universe.[9] Two state representatives, one in Georgia and one in Texas, recently defended geocentrism.[10] With the above exceptions, the geocentric model is gone—although humans still cling to ethnocentric and egocentric theories.

Then humans were demoted not just from being the center of the cosmos but from being the apex of life. Starting with the great nineteenth-century English naturalist Charles Darwin, scientists and then laypeople began to understand that humans were just one species of many. This is one of the reasons that creationists still attack evolutionary science. It is not so much the process of evolution they do not like but the way evolution has toppled humans from being lords of creation.

Perhaps the ultimate humbling of the human species has come from the studies of DNA. DNA not only encodes the instructions for life (the genes) but is also a record of what has happened in the past. Your cells contain not only the genetic instructions for making a human body but also old genes that the cells of your evolutionary ancestors used. These old instructions are crammed into your chromosomes like old memorabilia in an attic. Scientists have used DNA information, both new and old, to retrace the path of evolution of all life-forms. It turns out that almost all the genetic diversity is in bacteria, archaea, and protists. The fact that you may never have even heard of archaea and protists only goes to show you how much evolutionary history we humans are unaware of. (My software does not even recognize the word *protists*, and keeps changing it to *protests*.) The fungi, the plants, and all the animals—which we call the "higher" or "advanced" life-forms—are only a tiny subset of the variation within the protist group. There is about as much genetic difference between a bacterium in the soil and an archaean in a hot spring—even though they may look the same to us—than there is between a bacterium and you.

An ancient Israelite shepherd and king, David, looked into the sky and had feelings that most of us have had. He recorded his impressions in Psalm 8, in which he addressed God: "When I consider the heavens, the works of your hands, the moon and the stars, what is man?" Why should God care about anything as insignificant as man? If David had known just how big the cosmos was or how many creatures we share Earth with, he would have been even more humbled. But neither David nor the billions of people who have read his psalm have been content to leave humans in a position of humility. We have instead imagined a whole realm of angels, and that humans are merely a step or two below them. This was a "glory and honor," to use David's term, and was something that no quasar or glacier or dinosaur could have. David then proclaimed us to be the rulers of creation, which humans through the ages have taken as a license to trample everything. In contrast, the new scientific understanding of the place of humankind in the universe and among the life-forms of Earth means that we cannot look at the sky and Earth and think that they were made for us, or that our origin or demise means much of anything to the cosmos. Humans may still be significant in the eyes of God, but this significance exists entirely in a spiritual realm apart from the cosmos.

Humans: Marginal but Powerful

Astronomy and biology have therefore shown humans to be at the margin, not the center, of the cosmos and of life. But marginal as we are, we have proven to be a transformative power on Earth. As will be explained in the chapters of this book, humans have had an overwhelming impact on Earth, from the greenhouse effect to sexual selection to altruism to intelligence to evolution itself. Our species, especially within the last century, has put stress on the network of species and on the very life-support systems of the planet.

If Earth were conscious, she would look at us and see that individual acts of altruism continue, but that powerful technologies have greatly repressed altruism. Individual humans may want to create peace, but governments often want war. Individual humans may want to be unselfish, but unselfish individuals are beaten down by corporations. Individuals may want to reduce the devastating impact of the human species on the planet, but their efforts are swamped by collective greed and waste. Were she actually aware of us, Earth could be like a stressed-out, middle-aged mother of delinquent teenagers—teenagers who could severely harm her. They are eating her, chopping away at her, fouling her. They are certainly subjecting her to stress.

She is still beautiful, strikingly beautiful, as beautiful as she has ever been in four billion years. And we cannot kill her, for she will outlive our self-inflicted destruction and will do nothing to save us. However, we can disable her permanently. She runs the risk of entering her later years depleted of many of the species that made her youth so interesting, with a climate in which many of these species cannot survive. We humans have managed to understand the universe; to create spectacular works of art, literature, and music; and to perform astonishing acts of compassion, because of individual altruism and despite collective oppression. Earth will behold our individual virtues and collective evils until our species, like all others, becomes extinct.

Fundamentalists believe that humans are the masters of the world chosen by its Creator; that the few feet before we reach the rock pedestal in Kansas is the entire time that the universe has existed, and that great spiritual forces will soon bring it to an apocalyptic conclusion. They believe that the rest of the Continent of Time does not even exist. No wonder religious fundamentalists have so little trouble with the idea of human destruction of the planet; their myopic vision denies the existence of most of its history and its future. Even those of us who are not fundamentalists usually have a self-centered historical, economic, and personal focus.

Our self-focused attitudes must change if we are to avoid imposing irreparable damage on the living Earth. The focus of this book and of all of my work is to open people's eyes to the larger reality of space and time. We must use the best of our intelligence and altruism to be good caretakers of an Earth the quality of whose future depends on us.

Chapter 1

MEET MOTHER EARTH

This book is the biography of planet Earth, sometimes loosely called Mother Earth. "Mother Earth" is, of course, a figurative term. Earth is not a person and does not have a mind. The ancient Greeks considered Earth to be a goddess, Gaia, but modern science reveals that everything in the history and current operation of Earth has a physical, chemical, or biological explanation. There is no body or mind or goddess-spirit of Earth.

On the other hand, Earth is not just a stage upon which life acts or a storehouse from which life gets its resources. Earth is not merely the playroom and cafeteria of organisms. Twentieth-century ecologist G. Evelyn Hutchinson's phrase "evolutionary play in an ecological theater" suggests that the physical conditions of the Earth control evolution, but he also indicated that the Earth's ecosystems alter those conditions over evolutionary time.[1] Life forms a network of interaction. Usually the most important part of an organism's environment is other organisms. Moreover, this network of life has completely transformed Earth's temperature, its atmosphere, and even its geology. The activities of organisms have filled the air with oxygen, and even much of its limestone is biological in origin. The actors in the evolutionary play are making changes in the stage as they go along. It is as if the actors were the props.

Not only has the network of life transformed the conditions of Earth, but it appears to, in some ways and imperfectly, regulate those conditions, just as your body imperfectly regulates its temperature. Your body maintains a stable temperature because your brain stem sends out commands to your body to speed up or slow down processes that create or disperse heat. The Earth has no body or brain. It seems, however, to have a disembodied, mindless, but largely successful kind of self-regulation. This is why I have titled the book *Life* **of** *Earth*, not *Life* **on** *Earth*.

Many scientists have conceptualized this network of life as Gaia.[2] These scientists are not offering sacrifices at a Delphic Oracle or hallucinating on the carbon dioxide in the oracular caves. They are not nature worshippers. To them, Gaia is a personal name for a living planet that is not, or not quite, a person. Earth does not care for its organisms as would a mother; in fact, some scientists, such as planetary scientist Peter Ward, characterize Earth as being like Medea, rather than Gaia.[3] Medea, another character from Greek mythology, was the vengeful wife of Jason, whom she tricked into eating his own (and her) children. The network of life has often, like a good mother, enhanced the welfare of its inhabitants, but sometimes it has made conditions worse for them. Nevertheless we, and all other species, depend on this web of life as surely as any animal depends on its mother.

LUCKY GAIA

One day in 1950, over lunch with his scientific colleagues, physicist Enrico Fermi heard someone speculate about how many advanced civilizations there must be out in space. Fermi quipped, "So, where are they?" He meant that if there were many advanced civilizations, some of them must be more advanced than we are, they must have invented space travel, and at least some of them should have contacted us by now. This has come to be known as "Fermi's Paradox." One answer to this paradox is that there are so few advanced civilizations in the universe that they have not found us yet and probably never will. According to this view, sometimes called the "Lucky Gaia" hypothesis, Earth-like planets, Gaia planets, might be very rare. Very few planets have been as lucky as Earth.

It cannot be denied that, from the very beginning, Gaia has been very, very, very lucky. The network of life arose and developed to its present astounding complexity as a result of a vanishingly rare conjunction of circumstances presumably enjoyed by very few other planets in the universe.

Planets are not at all rare in the universe. More than four hundred planets have been detected orbiting nearby stars.[4] Astronomers can detect these planets and calculate their masses and orbit distances by the fluctuations in light intensity and by the slight swaying movements of the star itself. In one case, scientists were able to generate an image of the planet.[5] Most of these planets have been what astronomers call "hot Jupiters," that is, large gaseous planets orbiting very close to their stars. They resemble a double-star system in which the smaller star never ignited. But a few of the planets are believed to be rocky like Venus, Earth, and Mars. It is therefore likely that rocky planets are common in the universe.

Earth's luck goes far beyond just being a rocky planet, according to the Rare Earth hypothesis of planetary scientists Peter Ward and Donald Brownlee.[6] We can begin, Ward and Brownlee say, by thanking our lucky star, the sun.

Thank Our Lucky Star

First, the sun is a calm and stable star. Many stars, such as the Cepheid variable stars mentioned earlier, fluctuate wildly in their energy output: they are small and dim; then as quickly as one day to a couple of months later, they become larger and more than twice as luminous. Such pulsations in energy may prevent life from ever getting started on any planets that revolve around variable stars. In contrast, the sun has been stable for billions of years. Not perfectly stable, of course. The sun has had occasional "coronal mass ejections," in which it propels energy and particles from its outer layer out into the solar system. One of these mass ejections, on September 1, 1859, was strong enough that it shut down the telegraph systems in the United States and Europe and caused auroras to occur in the skies of even tropical regions. The sun also has an eleven-year sunspot cycle. These variations, however, have not had much effect on Earth. Coronal ejections have not been known to have ever harmed life on Earth (no organisms were harmed when the telegraphs shut down), and solar intensity shifts by only 0.1 percent during the sunspot cycle. The sun has, in fact, changed its energy output over the billions of years of its existence. It has increased the intensity of its radiation by about 30 percent during that time—but it has done so very gradually.

Second, the sun is an isolated star. Many stars have partners, forming multiple-star systems—most commonly, binary systems in which two stars swing around each other like dancers.[7] If the sun were part of such a close family of stars, the other stars would prevent planets from having stable orbits, which might prevent the evolution of life. The sun is also far away from stars that emit so much energy that they would disrupt or destroy life. For example, a supernova anywhere within a few dozen light years of Earth would wipe out all life—but there have been no supernovae in the sun's neighborhood for at least several billion years. Moreover, stars in the centers of galaxies may be so close together that they would disrupt the revolution of one another's planets, even if they are not part of multiple-star systems. But the sun is on a swirling arm far from the center of the galaxy. If we were near the center of the galaxy, many stars would be so close to us that night would not be very dark, and those stars would yank and tug us around and disrupt the stability of our planet's conditions.

Thank God for Jupiter

Ward and Brownlee also point out that Earth resides in a very lucky neighborhood of the solar system. The two sources of luck are Jupiter and the moon. First, consider Jupiter.

When the solar system first formed, it was a disc of small asteroids. Many of these asteroids ran into each other and were crushed into planets by their own gravity. These planets continued to mop up asteroids until about 3.9 billion years ago. After that time, few asteroids remained that could crash into the planets. Most of the craters on the moon (which, as large as some planets, also helped to clear away asteroids) are older than 3.9 billion years. One rare and visible exception is Tycho, the large crater in the moon's southern hemisphere with white radiating ejection plumes visible through amateur telescopes. Tycho was formed by an asteroid that hit the moon only 108 million years ago. The moon, which has no wind or weather, has preserved an intact sample of the asteroid impacts that imperiled the early solar system.

Another important component of the solar system is comets. There are billions of these dirty balls of ice that orbit the sun just beyond the outer edge of the solar system. Most of them remain at the edge of the solar system, but some of them have very elliptical orbits, which bring them close to the sun. They whip around the sun like a slingshot and fly back out into the outer edges of the solar system. While comets are near the sun, solar radiation vaporizes some of the water, creating the comet's "tail" that everyone recognizes. Before 3.9 billion years ago, there were a lot of comets, but they are now, like asteroids, comparatively rare.

The principal reason that asteroids and comets now only rarely fall from the sky is the planet Jupiter. Jupiter is so massive, and has such a powerful gravitational field, that it has sucked up most of the asteroids in the inner solar system, except for those in the asteroid belt, whose orbits have been stabilized by that same Jovian gravitation. Any asteroid or comet that happens to come within several million miles of Jupiter is drawn inevitably into its gaseous embrace. This is exactly what happened to the comet Shoemaker-Levy 9 in 1994. After whipping around the sun and heading back into the outer reaches of the solar system, this comet slipped too close to Jupiter, whose gravity fractured it into pieces. Each piece created a huge flare of radiation as it fell into Jupiter's dense atmosphere, and each of the black spots that remained visible for a few weeks was similar in size to the Earth. Therefore Jupiter continues to clear away asteroids and comets from the solar system. Without Jupiter, asteroids and comets might be hitting Earth so frequently that life would not have a chance to exist for very long.

The Moon—It's Not Just Pretty

And then there is Earth's closest neighbor, which poet Percy Bysshe Shelley (in "The Cloud") described as "That orbed maiden, with white fire laden, whom mortals call the moon." Most planets have moons, but Earth is the only planet in the solar system with a moon so large in relation to it. Mars has two tiny moons, Deimos and Phobos, named after the two horses of the war god's chariot. Jupiter and Saturn have moons larger than ours, but they are tiny in relation to the planetary masses. Our moon is large enough and just far enough away to profoundly influence our planet without severely disrupting it. Everyone knows that the tug of the moon causes the tides. Were it not for tides, there would be no intertidal zone, the only home of thousands of species of organisms. But tides may be of relatively little importance to the planet as a whole, even though they are important to barnacles. The major effect of the moon on Earth, crucial to the survival of life as a whole, is to stabilize Earth's movement.

As planets revolve around their suns, they rotate on their axes. These rotational axes wobble, pointing in different directions at different times. Any planet with a large amount of wobbling would have unstable climatic zones, since sometimes the equatorial zone and sometimes the polar zones would directly face the sun. The part of a planet directly facing its sun will receive the most intense radiation and will be warmest. How could tropical, temperate, and polar plants and animals evolve if the climates of those zones are extremely variable? This appears to have happened with Earth's less fortunate little brother, Mars. Earth, however, has not tilted more than about twenty degrees from the plane of its revolution. Even the little bit of wobbling that Earth does experience has been enough to cause about twenty ice ages during the last two million years of Earth's history. We have the moon to thank for the relative stability of Earth's movements.

The Face of the Earth

The surface of Earth consists of restless plates separated by fractures or faults. The plates are always moving, propelled by rivers and plumes of lava beneath them. Two plates might move apart, with lava erupting from the fracture and creating new plate surface. This is what is happening along the Mid-Atlantic Ridge, which looks like a scar down the middle of the Atlantic Ocean. The plates' movement is making the Atlantic Ocean wider at about the same rate that fingernails grow. Two plates might ram into each other, with one of them slipping underneath the other. This is what is happening in the Marianas Trench near Indonesia. Two plates can scrape against each other, as along the San Andreas Fault in California. Earthquakes and volcanoes are common in all regions where two plates move apart, crush together, or scrape. Because of these movements, the rocks of these plates are continually recycled and renewed. There is probably no part of the ocean floor that is more than 120 million years old.

Meanwhile, continents (which consist of lighter minerals than the plates themselves) sit on the tops of the plates and are tossed around by them. As the plates move, lighter minerals rise to the top, and over billions of years, these lighter minerals have formed the continents. Because they sit on top of restless plates, the continents are also pulled apart and crushed back together. About a billion years ago, the supercontinent called Rodinia was pulled apart into separate continents, which were then crushed back together into a supercontinent called Pangaea about a quarter of a billion years ago. Pangaea was then pulled apart into the modern continents. Continents do not get recycled beneath the plates. Unlike ocean floor plates, continents can be very old.

The oldest continental areas are about 3.8 billion years old. There are, therefore, no continental areas that preserve a record of the heavy asteroid bombardment that occurred about 3.9 billion years ago. Asteroids have been rare during the last 3.8 billion years but common enough to leave a few dozen craters on the continents, from the huge two-billion-year-old Vredefort Crater in South Africa to the small fifty-thousand-year-old Barringer Crater near Interstate 40 in Arizona (figure 1-1).

Figure 1-1. Barringer Crater in Arizona resulted from a meteor impact about 50,000 years ago. Meteor impacts are now rare on Earth but were very common prior to 3.9 billion years ago. Image courtesy of United States Geological Survey.

Goldilocks's Earth

Finally, the Earth itself is a mighty lucky planet. It happened to form in the "habitable zone" of the new solar system, in which temperatures were in the right range to allow water to exist in liquid form. No medium other than water is known in which lifelike processes can occur. Some scientists speculate that the liquid methane on Saturn's moon Titan may be a medium for life. Even if this is the case, molecules move very slowly in liquid methane, and any metabolism of life-forms on Titan would be very slow, and the resulting life-forms would be very simple. And Earth has plenty of water. The water may have been delivered to the young Earth by comets hitting it before 3.9 billion years ago. The ice in the comets melted and vaporized, creating a haze of steam, much of which was lost into space as new comets continued to rain from the sky. When the collisions became less frequent, and Earth cooled down, the steam became oceans, and water vapor saturated a hot, dense atmosphere of carbon dioxide and nitrogen gases. Mars, Earth's little brother, also had oceans when it was a young planet.

The Earth is also just the right size. If the Earth were too large, its gravity would be so great that complex organisms (not to mention mountains or even continents) would not be able to stand up. If the Earth were too small, however, it would be unable to hold onto its atmosphere, partly due to a lack of sufficient gravity, and also because particles streaming from the sun would have stripped the gases away. Mars, which is about half the size of Earth, has an atmosphere—just barely. Its atmosphere is about 1 percent as thick as ours. At first this seems strange, given that its gravity is at least one-quarter as strong as that of the Earth. But Mars is small enough that its core has cooled off and solidified. On Earth, the currents of molten lava produce a magnetic field that deflects much of the dangerous solar particle stream, except for rare events such as the solar flare of September 1, 1859. Mars has no such protection. The solar particles have scoured away most of its atmosphere, as well as its surface water. When Earth and Mars first formed, they were both wet planets with carbon dioxide atmospheres. Earth kept its atmosphere and evolved; Mars lost its atmosphere and (apparently) died. The surface of Mars is without life; and if there is life on Mars, it is deep in the rocks and therefore microbial in size.

The fact that the Earth is not too large and not too small, and is just the right distance from the sun, has been compared by some scientists to the story of Goldilocks, the Aryan girl who thought she had the right to barge into somebody else's house. After sampling various portions of food and sizes of beds, she found that Baby Bear's food and bed were "just right." And if Earth were not "just right" in size and chemical composition, there would be no life on Earth more complex than microbes.

Our planet is also fortunate to have radioactive elements in its core. Without radioactivity, the core of the Earth might have cooled down and solidified, causing Earth to have a fate similar to that of Mars, though it would not have happened quite as quickly.

Our planet also happens to have plenty of carbon, which may be the only element from which life can be built and sustained. As any aficionado of *Star Trek* knows, silicon-based life-forms such as the Horta are conceivable, since silicon atoms have several chemical similarities to carbon. However, silicon is probably too heavy to allow a silicon-based life-form to participate in a silicon-based ecosystem. Silicon is a mineral; it is always a mineral. Carbon atoms can form carbon dioxide gas, which circulates through the atmosphere and is turned into complex molecules by photosynthesis. But silicon dioxide is quartz and just stays in the crust of any planet on which it is found.

The most obvious way in which Earth is lucky is that it has water—lots of it. Earth is not quite unique in this respect—Jupiter's moon Europa is covered with oceans that are capped with ice. But on Earth, the water exists in all three states: ice, liquid water, and water vapor. Not only do life processes, as we know or imagine them, occur in liquid water, but water moves from oceans to continents and back in gaseous form. Europa has liquid water (kept from freezing not by any warmth from the sun but by the pull of Jupiter's gravitational field), but it has no water cycle, as does the Earth. Thus when you consider carbon and water, Earth is lucky not just in what it has but in what it can do with it: carbon and water can circulate around and around on the planet.

Putting all these facts together, Ward and Brownlee conclude that simple microbial life may be common in the universe. But for advanced life to evolve, it is necessary that planetary conditions remain within certain limits for a long time. Such long-term stability, and the advanced civilizations that would require such stability, appear to be vanishingly rare in the universe. This is one answer to Fermi's Paradox.

Bad Luck

Earth has also had its share of bad luck. Some of the bad luck has come from those few asteroids and comets that do happen to collide with Earth. The most famous example is the asteroid that hit Earth sixty-five million years ago (the end of the Cretaceous period) at what is now the tip of the Yucatán Peninsula.[8] This impact created Chicxulub Crater, which has been largely obscured by erosion. The asteroid created a firestorm as forests burned, not only from its direct impact but from red-hot blobs of lava that were ejected and then fell back to Earth, creating firestorms in new places. The smoke obscured sunlight worldwide for several years afterward. Up to 50 percent of the species on the planet became extinct. Among the organisms that became extinct were all the dinosaurs (except birds, which are feathered dinosaurs). It was this event that opened up the opportunity for mammals to evolve into so many forms. Mammals had existed prior to the Chicxulub event for even longer than they have existed since, but large mammals could not compete with large dinosaurs. Most mammals were small and nocturnal, no more of a threat to dinosaurs than to eat an occasional hatchling.[9] The asteroid was permanent bad luck for dinosaurs and temporary bad luck for the ecosystems of Earth as a whole; the aftermath was, for mammals, at least, good luck.

Some of the bad luck in Earth's history arises from volcanic eruptions. A single volcano can alter the climate of Earth. This concept was expressed by Benjamin Franklin. While he was the American ambassador to France in 1783, Franklin noticed that the weather was unusually cold, and he speculated that the cold weather was caused by the eruption of Laki, a volcano in Iceland. He was right. The eruption of the Indonesian volcano Tambora in 1815 put enough dust and sulfur dioxide into the atmosphere, which blocked enough sunlight to cause agricultural failures in Europe and North America; 1816 was called "the year without a summer." (It was during the non-summer of 1816 that Percy Bysshe Shelley and Mary Wollstonecraft Godwin, the woman he later married, stayed in Lord Byron's Switzerland villa. Since the weather was so dreary, they decided to write horror stories. Mary Shelley's story became the novel *Frankenstein.*) The 1991 eruption of the Philippine volcano Pinatubo projected dust and sulfur dioxide into the air, which reflected enough sunlight to lower Earth's average temperature by about a third of a degree.

But these events are nothing compared to what can happen when a whole bunch of volcanic explosions (a "supervolcano") occur. Two hundred fifty million years ago, at the end of the Permian period, volcanoes in what is now Siberia erupted for about two hundred thousand years, producing layer after layer of lava; the deposits are called the Siberian Traps. The sulfur dioxide from these eruptions caused acid rain, and the carbon dioxide caused global warming, extensive enough to have at least initiated the greatest mass extinction in Earth's history: the Permian extinction wiped out about 95 percent of species. Biodiversity recovered after this event, but the recovery took almost one hundred million years.[10]

Both the Permian and Cretaceous extinctions had decisive effects on the direction of evolution on Earth, and both of them were chance events. As evolutionary scientist Stephen Jay Gould pointed out, brachiopods such as lampshells, and bivalve mollusks such as clams are similar to one another; one does not appear to be superior to the other at living in marine environments. But, by chance, lampshells suffered more from the Permian extinction than did bivalves. It may therefore have been the Permian extinction that made lampshells rare and allowed clams to become common enough that, as Gould said, "Ho Jo can feed a nation on their breaded feet."[11]

There is, of course, a possibility that supervolcanoes and asteroids can wreak havoc on Earth again. In 1908, a comet almost hit Siberia. It vaporized before actually hitting the ground, but it knocked over trees for hundreds of miles around and created a dust cloud that reflected sunlight all the way to Western Europe. Had the comet struck over a densely populated area, the damage would have made it one of the major events in the world. It was only the sheer luck of ramming into the air above Siberia that made the Tunguska Event a footnote to human history.[12]

A handful of astronomers are charting the courses of "near-Earth asteroids" whose paths cross Earth's orbit. No known asteroids will come close enough to Earth to endanger it on any timescale meaningful to us. Perhaps the greatest danger is from the asteroid Eros, which has a 50 percent chance of hitting Earth during the next hundred million years. The "near misses" predicted for 2029 and 2036 will not come within millions of miles of our planet. Of course, asteroids are small, and there are a lot of them we do not know about. While no known comets are on a collision course with Earth, there are millions of them surrounding the solar system, and we never know when a new one will come flying toward the sun. If an asteroid or comet threatens Earth, we will not be able to send up space cowboys to blow it up, as in the 1998 Bruce Willis movie *Armageddon*. Blowing it up would only transform a big asteroid into lots of little ones, all on an essentially unchanged course toward Earth. The astronomers can legitimately wonder whether a world of Republicans and Democrats and Sunnis and Shia, or their twenty-fifth-century equivalents, would be able to put their differences aside long enough to deflect an asteroid or comet from its deadly path.

Lucky Life

The origin of life itself was a lucky accident. When the oceans formed on an Earth no longer molten, conditions were hellish. In fact, scientists call this the Hadean period, named for Hades, the Greek god of the underworld. Within a few hundred million years, these oceans were filled with living cells that resembled modern bacteria. The quick emergence of life may have been the kind of sheer luck that occurred on only a very few planets.

Scientists have not yet figured out how the nonliving molecules of the primordial Earth became the molecules of life, or how these molecules assembled into cells.[13] In 1953, Stanley Miller, a PhD student, became the most famous biochemist in the world when he conducted the first simulation of the origin of life. He mixed together gases and water and used an electrical spark to make them into organic molecules. The result of his experiment was mostly amino acids, the building blocks of proteins, which are some of the essential molecules of life. This one experiment demonstrated that random chemical reactions could produce some of the molecules from which life is made. In the half century since that experiment, hundreds of chemists have tried to figure out the steps by which the molecules of life, and the first cells, might have originated. (I say "molecules of life" rather than "organic molecules," because many organic molecules such as benzene are not involved in the chemical reactions of living cells. I also avoid the term "living molecules" because the molecules are not, by themselves, alive.) They have not yet been successful, but every time the quest seemed to be impossible, a new breakthrough occurred—just enough to keep the scientists from giving up.[14] Apparently, there was nothing inevitable about these chemical reactions. The first simple living cells came into existence by luck.

Many people think of evolution, and all of life, as sheer luck. Of all the planets in the universe, ours just got lucky. That is part of the truth. But luck is not the whole story.

GAIA: MORE THAN JUST LUCK

As soon as simple cells became abundant on Earth, they began to change the planet. They were major components of one another's environments. They evolved in response to one another. As will be explained later in this book, some of these simple cells fused together to form more complex cells. Some of these complex cells evolved into multicellular organisms. From the very beginning, and ever more with the passing of time, life began to form a web, or network, of interactions. The story of evolution is not just about organisms responding to the physical conditions—for example, temperature and chemistry—of Earth, but is about organisms evolving within this living, planetary-scale framework that some scientists call Gaia. Gaia started by sheer luck, but once Gaia was born, its life was not just a matter of luck. In many ways, it took care of itself and grew itself into the living biosphere that it is today. It has helped to create the conditions that keep it alive.

In particular, the biosphere as a whole appears to regulate itself and stabilize its own conditions. Examples include temperature and atmospheric composition. The temperature of Earth, and the amount of oxygen and carbon dioxide in its atmosphere, appears to be regulated, such that it does not fluctuate as much as would be expected on a lifeless planet. This is the idea behind the Gaia Hypothesis that was first proposed by atmospheric scientist James E. Lovelock and microbiologist Lynn Margulis.[15] Since Gaia has no mind and no genes, it cannot control itself the way an organism does. Instead, the apparent regulation of temperature and atmosphere is an emergent property of the activities of living organisms.

Emergent Gaia

An emergent property is a seemingly complex process that results from the operation of simpler processes. One example of an emergent property on Earth today is the collective intelligence of a colony of ants. Ants appear to have a sinister and frightening intelligence. When they invade your house, they seem to be able to find anything they want. Once I left a bag of potato chips, the top curled shut but not ant-proof, on the top of my refrigerator. One tiny tip of the bag touched the wall. When I came home, streams of ants were heading from my floor into the bag. How could they have figured out how to do this? The answer is that no single ant has very much intelligence. Each ant is like a little robot that just randomly searches its environment for food. Once one of these ants finds food, it returns to the colony. As it does so, it lays a scent trail. Other ants, detecting this scent trail, abandon their random searches and follow the trail. Before long, dozens of ants are following the trail to the food. They, in turn, lay down scent trails, creating a scent superhighway. Soon thousands of ants are traveling on this superhighway. All an ant has to do is search randomly, without laying down a scent trail, lay down a scent trail if it finds food, or stop searching if it detects another ant's scent trail. An ant's job is not very complex. But when thousands of ants follow this simple protocol, the result is an astonishingly complex foraging system. The system emerges from the simple activities of the ants. Nobody is in charge, and nobody tells the ants where to go. Complexity emerges from simplicity.[16]

Gaia's planetary-scale regulation of temperature, oxygen, and carbon dioxide is also an emergent property, resulting from the physiological processes of its component organisms. Two of these processes are respiration and photosynthesis.

Respiration and Photosynthesis

Cells ingest food molecules and break them down into simple waste products. *Respiration* is the way cells break sugar down into carbon dioxide. They do this in order to release energy from the sugar, energy that the cells can use for their own chemical reactions. This process forms electric currents, which deliver energy to the cellular reactions. Once the electrons have lost most of their energy, they need someplace to go. Most organisms direct these worn-out electrons into oxygen gas, turning it into water. Respiration, therefore, releases energy from food, produces carbon dioxide, and consumes oxygen. Cells have been removing oxygen from and releasing carbon dioxide into the atmosphere for billions of years.

Photosynthesis occurs when pigments (usually chlorophyll) absorb sunlight and use it to create an electric current. The energy comes from the sunlight itself. The electrons of the current come from the splitting of water molecules, resulting in leftover oxygen molecules. The electric current then provides energy and electrons that allow carbon dioxide molecules to be built up into sugar and, eventually, into all the complex molecules of a cell. Photosynthesis, therefore, puts oxygen into the atmosphere, removes carbon dioxide from the atmosphere, and builds complex molecules. Billions of photosynthesizing cells can clear carbon dioxide out of the atmosphere and fill it with oxygen. And this is exactly what photosynthetic cells have done.

At the time life began, the atmosphere probably had a lot of carbon dioxide in it. Carbon dioxide creates the famous "greenhouse effect." It does this by absorbing infrared radiation (invisible light, just beyond the red end of the visible spectrum) that would otherwise be lost into outer space. Earth was much warmer 3.8 billion years ago than it is today. Its average temperature was about 50 degrees C, which is about 122 degrees F—and this, despite the fact that the sun was 30 percent dimmer at that time. Today, even though the sun is brighter, the average Earth temperature is about 13 degrees C (about 55 degrees F). The rough parity between greenhouse effect and solar intensity was the equivalent of what today would be several million miles of distance from the sun. That is, life itself has expanded the "habitable zone" of the solar system. In order to remain in the right temperature range during the last four billion years, Earth would have had to move several million miles farther away from the sun, an obvious impossibility. But the processes of Gaia, primarily photosynthesis, have accomplished an equivalent miracle.

The history of life has largely been the story of respiration and photosynthesis. Photosynthesis makes organic food, and respiration releases the energy from it. Billions of years ago, photosynthetic bacteria made a living through photosynthesis, and respiratory bacteria survived by respiration. Ever since that time, complex photosynthetic cells, and the plants that evolved from them, have made a living through photosynthesis, and other complex cells, and the animals and fungi that evolved from them, have made a living through respiration. (Plant cells also have respiration, but less of it than photosynthesis.) Over billions of years, photosynthesis has put oxygen into the air and removed carbon dioxide from the air, but not too much, and respiration has done the opposite. Photosynthesis has removed much of the carbon dioxide from the air, transforming it into organic compounds, both living and dead. Photosynthesis, therefore, brought the period of extreme hothouse conditions to an end. Photosynthesis also filled the atmosphere with oxygen. The accumulation of oxygen was gradual, and oxygen did not reach its present level of abundance until about six hundred million years ago.

Photosynthesis is not the only process that removes carbon dioxide from the air. As the continents arose from the oceans, minerals in the exposed rocks reacted with carbon dioxide and created inorganic limestone. Much of the carbon dioxide in the primordial atmosphere became inorganic limestone. At the same time, many microscopic photosynthetic organisms have carbonate shells. As these organisms die, they settle to the bottom of the sea, causing a buildup that eventually becomes organic limestone. Organisms have, in this as in other ways, changed the very geology of the planet. Some of the limestone in the crust of the Earth—and there is a lot of it—was produced by photosynthetic organisms or by organisms that ate them.

Lucky Balance

For almost a billion years, photosynthesis and respiration have been roughly in balance. As a result, both oxygen and carbon dioxide contents of the atmosphere have remained within broad limits that life can tolerate.

Today, the atmosphere contains 21 percent oxygen. About three hundred million years ago, the air may have had as much as 35 percent oxygen. That is almost enough oxygen to cause all the organic matter and organisms in the world to spontaneously combust.[17] Then, about 250 million years ago, the atmospheric oxygen content may have plummeted to about 15 percent, low enough that many vertebrates could not breathe very well, and none of them could have breathed at all at elevations higher than a few thousand feet.[18] This cannot be called fine-tuning—the range between 15 and 35 percent is pretty severe—but oxygen content has not gone close to zero or high enough to literally burn everything up.

About three hundred million years ago, there was almost ten times as much carbon dioxide in the air as today, at least three thousand parts per million (0.3 percent), creating a worldwide hothouse, though not nearly as much as the Earth experienced before the evolution of life. The lowest concentration of carbon dioxide has been during the most recent ice ages, when the air had less than two hundred parts per million (0.02 percent) carbon dioxide. This cannot be called fine-tuning—the range between 3,000 ppm and 200 ppm is pretty severe—but there has always been enough carbon dioxide to keep the Earth warm enough for life, but not too much as to bake it to death.

This regulation of oxygen and carbon dioxide levels is not sheer luck but is the result of negative feedback. Negative feedback occurs when one process limits another. When there is a lot of carbon dioxide, the rate of photosynthesis increases and removes more carbon dioxide from the atmosphere. When there is a lot of oxygen, the rate of respiration increases and removes more oxygen from the atmosphere. That is, photosynthesis limits respiratory carbon dioxide buildup, and respiration limits photosynthetic oxygen buildup. What photosynthetic organisms consider to be waste (oxygen) is a resource to respiratory organisms, and vice versa. Some scientists say that this is how Gaia regulates Earth's atmospheric composition. Planetary scientist Tyler Volk calls Gaia "Wasteworld" for this reason.[19]

"Wasteworld" can operate on more local levels as well. As I will explain in chapter 4, bacteria live in the roots of many plants. These bacteria transform atmospheric nitrogen into nitrogen fertilizer (ammonia). These plants therefore have their own built-in bacterial fertilizer factories. But it is expensive: the plants have to feed the bacteria. If the soil is depleted in nitrogen fertilizer (ammonia, nitrite, and nitrate), the cost is worth it to the plant. But if the soil already has a lot of nitrogen in it, the cost of this bacterial "nitrogen fixation" is not worth it. Depleted soils create conditions that benefit nitrogen fixation. The plants grow and then die; the decomposers release the nitrogen from the leaf litter, enriching the soil with nitrogen. Soon the soil is no longer depleted, and plants that do not have bacterial nitrogen fixation become more common. Some plants capitalize on the nitrogenous wastes of other plants (via decomposers) while other plants capitalize on the nitrogenous wastes of bacteria that live in their roots.

From this viewpoint, Gaia, once again, just got lucky. Negative feedback helps but is no guarantee of success. Photosynthesis kept the greenhouse effect from going out of control, but what if there had been just a little bit less photosynthesis? The greenhouse effect might have gone out of control. Suppose there had been just a little bit too much photosynthesis? There might have been too much oxygen in the air. The balance of carbon dioxide, oxygen, and temperature is not simply due to luck; but the *success* of negative feedback in keeping Earth conditions within the limits required by advanced life-forms may have been sheer luck.

After all, nature also has processes of positive feedback, in which one process enhances another. Two examples are allelopathy and fire.

Some plants, such as walnut trees and the shrubs that live in arid zones, release poisons into the soil. These poisons inhibit the growth of other plants, but not of the plants that produce them. The more poison that is produced, the more land area is available for those plants to grow. More poison means more walnuts, which means more poison, and so on. Perhaps it is just luck that walnut trees can grow only in a restricted range of temperature and moisture conditions; outside of that range, the poison does them no good.

Some plants are adapted to a fire cycle. To them, fire is a blessing: it burns up dead stems and leaves, releasing the nutrients back into the soil from them. Some of these plants, such as prairie grasses and wildflowers, as well as many oak trees, can resprout quickly after a fire from buds that were protected down in the soil. Other plants, such as certain species of pines in coastal California, eucalypts in Australia, and proteas in South Africa, release their seeds only after fires. Within a year after the fire, the next generation of these plants is growing vigorously. These plants actually encourage the spread of fire. Chaparral shrubs, for example, accumulate dead stems and leaves, and even produce combustible chemicals in their wood (on a warm day, you can smell the creosote-like scent). The more fire, the more land area for these plants to grow, which means more fire, and so on. Perhaps it is just luck that these plants are limited by soil and climate conditions, and therefore do not turn the whole world into a tinderbox. Actually, about a quarter of the Earth's land area was originally covered by grasslands, most of which depend on a fire cycle that burns away most of the woody plants.

Perhaps Gaia is just lucky that the negative feedback processes are greater than the positive feedback processes. When the atmosphere had 35 percent oxygen, it came pretty close to burning up the Earth. At that time, much of the land surface was covered with swamps. But even some of the fossils from these swamps show evidence of burning. Perhaps Earth survived this period, the Carboniferous period, by the skin of its teeth. An abundance of plants and animals lived on Earth at that time, oblivious to the fiery inferno they were narrowly missing.[20] When the atmosphere had 15 percent oxygen, at the end of the Permian period, life was at risk. And not just for that reason. The Permian extinction was the biggest catastrophe for life in the last 550 million years. Up to 95 percent of all species became extinct. Perhaps Gaia survived this time, also, by the skin of its teeth. Vertebrates narrowly survived. Just after the Permian extinction, during the early Triassic period, most of the vertebrates running around on the land were of just one species—*Lystrosaurus*, a generalist pig-like reptile that could and did eat whatever it found lying around.[21]

GAIA GROWS UP

But the history of Earth has not simply been the good luck conferred by negative feedback processes. The evidence for this comes from looking back further into the past, before 550 million years ago, back to the time when most organisms on the Earth were microscopic, back when Gaia was still too young and small to have much of a regulatory effect on the conditions of the planet.

Snowball Earth

Between three billion years and one-half billion years ago, the Earth experienced three periods of glaciation. These were no normal periods of glaciation; they were worldwide. Glaciers extended down even into the tropical regions. The evidence for this is found in the piles of glacial rubble discovered at the edges of continents that were at that time in tropical latitudes. Scientists conclude that the Earth, during those three periods of time, was almost completely covered in ice. The popular term for these periods is Snowball Earth.[22]

You might expect that sunlight would eventually melt the ice, and, in fact, it did. But ice reflects most of the sunlight right back into outer space, without absorbing it. This reflected light does not warm up or melt the ice, which is why the sun can shine on white snow all day without melting it very much. Dark objects, on the other hand, absorb sunlight and become warm. This is why, after a snowstorm, most of the melting occurs near dark objects such as tree trunks. Dead leaves on the snow absorb sunlight and melt their way down into the snow. There is a critical point—nobody is quite sure what that point is—at which the Earth would have been covered with so much ice that the balance would have been tipped into eternal snowball conditions. Earth was very lucky to not have gotten beyond that point, even though it had three chances to do so.

The ice on Snowball Earth did not cover volcanoes. Almost the only non-white surface of the Earth may have been little circles around volcanoes, which spewed carbon dioxide into the air. A strange situation may have resulted. The atmosphere might have been filled with carbon dioxide, enough to cause an intense greenhouse effect, only the greenhouse effect was prevented because the ice reflected the sunlight. The Earth produced little infrared radiation. A huge amount of carbon dioxide surrounded a frozen Earth. Then one day, or at least over a relatively brief period, enough of the ice melted that the balance was tipped the other way. Once there was enough dark ground to absorb sunlight and emit infrared radiation, the carbon dioxide had something to absorb. The greenhouse effect kicked in with a vengeance, melting the glaciers rapidly. The high concentration of carbon dioxide in the air could now react with the minerals in the rocks, creating inorganic limestone. Indeed, a layer of inorganic limestone sits right on top of the glacial rubble in the geological deposits from these periods of Snowball Earth.

Snowball Earth—Never Again?

The early Earth experienced really wild swings of temperature—very hot and very cold. And it might have been Gaia (acting more like Medea, in this case) that created the problem in the first place. Photosynthetic microbes might have used up so much of the carbon dioxide that there was insufficient greenhouse effect to prevent the Earth from freezing. And Gaia might have helped to get the Earth out of its predicament. There is some evidence that algae-like microbes may have grown under thin layers of ice, absorbed sunlight, and helped to melt the ice.[23] Regardless of these possibilities, the Earth experienced wilder fluctuations before 550 million years ago than it has since.

Before 550 million years ago, Gaia was in its infancy, so to speak. There were only microbes, and these microbes lived only in the oceans. The negative feedback processes of photosynthesis and respiration were operating, but not very much: the total biomass of life was not very great. Gaia was small and thin on the Earth. Starting about 550 million years ago, life exploded in diversity and in sheer tonnage. By the time the atmosphere had 35 percent oxygen (300 million years ago), or 15 percent (250 million years ago), there was enough photosynthesis and respiration going on to prevent a repeat of either primordial greenhouse effect or Snowball Earth conditions. Gaia had grown up, in both size and complexity. This suggests that the rough equilibrium of Earth conditions exists not just because Gaia got lucky. As Gaia has grown, the Earth has become more stable. In the last two million years, the Northern Hemisphere has experienced about twenty periods of glaciation, four of which were severe. These were the ice ages of which most people have heard. Over the northern United States, there was a two-mile-thick layer of ice. Some scientists speculate that it is because of Gaia, the network of photosynthesis and respiration, that these ice ages never became severe enough to create a new Snowball Earth.

Can Gaia Save Us?

In conclusion, we see that Gaia will probably stabilize the conditions of the Earth just enough to prevent life from being wiped out. Earth will probably never brush as close to death ever again as it did during Snowball Earth times, until the sun's final expansion. But this does not mean that there will never be any dangers.

In particular, the human economy depends entirely on not just a general but a very specific stability of Earth's climatic conditions. Global warming is the major crisis of the Earth today, not because it threatens life with extinction, but because it threatens to change the conditions of Earth just a little bit—just enough to disrupt the human economy. Even the wildest predictions of global warming do not far exceed about five degrees C of warming. As we have seen, the Earth has experienced fluctuations larger than this in the past. Global warming will not make us all fall over dead.

But it doesn't have to. The human economy is already on the brink. People are crowded in poor countries, and there is just enough food to feed most of them. It would not take much drought to reduce agricultural production enough to cause famines. It would not take very much of a rise in sea levels to flood the homes of millions of people and cause them to flee inland, across national borders, resulting in wars. Just a slight change, less than the changes that occurred before and after each of the twenty recent ice ages, would send our crowded, economically interdependent Earth into crisis. Many species would become extinct, but most would not. Humans would not. But remember that the modern human environment is not the outdoors but the built environment of houses, workplaces, and supermarkets. During the upcoming greenhouse crisis, you will be able to walk outside without finding it much warmer than it is now. But this will provide little comfort if you cannot get food because of economic and political disruption (see chapter 9). Gaia may save us from a severe greenhouse effect, but not from one that is big enough—and that will come soon enough—to severely disrupt human life.

BACK TO FERMI'S PARADOX

This brings us back to Fermi's Paradox. As astronomer Carl Sagan thought about what Fermi said, he began to be alarmed. His interpretation of this paradox was that advanced civilizations should, in fact, be common, and there should be a lot of them trying to contact us. That none seem to be out there implies that none of them developed long enough to have advanced space-traveling civilizations. This could only mean that advanced civilizations destroy themselves before they get that far, a viewpoint Sagan published in 1966.[24] This was during the cold war, when there was a very real possibility that the major civilizations of the Earth would blow themselves up in a nuclear holocaust. It was at that point that Sagan became an activist against nuclear proliferation and in support of environmental issues. He even spearheaded a briefly famous idea in 1983 that nuclear war would create so much smoke that it would plunge the Earth into a new ice age (nobody at that time yet knew about Snowball Earth) from which it might not emerge, a possibility that he and his colleagues called Nuclear Winter.[25] Sagan's first wife, Lynn Margulis, was one of the principal architects of the Gaia Hypothesis. Put their viewpoints together, and the conclusion you would reach would be that nuclear war could have a significant enough effect that it could even kill Gaia.

Since 1983, it has become clear that even a full-scale nuclear war (which, since the collapse of the Soviet Union, is much less likely) would create, at most, a nuclear autumn, a chill rather than a deep freeze. Recent evidence

Chapter 2

INEVITABLE EVOLUTION

THERE'S NO WAY TO STOP IT

There is no way to stop evolution.

By about three and a half billion years ago, Earth was covered with oceans. Microscopic cells filled these oceans and formed a self-regulating network that some scientists call Gaia. Over time, this network of life evolved into an ever-more complex system. As explained in the previous chapter, the self-regulation may have been the result of sheer luck. But the evolution that occurred within the Gaia network was not luck at all. Once life existed, evolution was inevitable.

Cats, Catalysts, and the Life of the Cell

You may be nothing more than a complex set of chemical reactions. Solar energy raises the electrons of atoms to high levels of energy in photosynthesis, and all other metabolic reactions allow this energy to flow to background levels. As biochemist Albert Szent-Györgyi said, "Life is nothing but an electron trying to find a place to rest."

Everything that happens within cells (and viruses also, which are not cells) is controlled by enzymes. Every chemical reaction in your body occurs because an enzyme enables it. Most enzymes are lumpy proteins that have little grooves in them into which molecules can slip and react. The shape of the groove is so specific that, in most cases, each kind of enzyme controls just one kind of chemical reaction. The enzyme-controlled chemical reactions determine which structures the cells should have and how they should live: they build everything from membranes to bones.

In some of these reactions, such as those involved in digesting food, enzymes break down large molecules into smaller ones. In other reactions, enzymes build up smaller molecules into larger ones. The reactions that break down large molecules into small ones do not require an input of energy, but the reactions that build up large molecules require an input of energy. This is consistent with the laws of thermodynamics, which say that large, orderly molecules tend to break down into small, disorderly molecules. Small, disorderly molecules can be made into large, orderly molecules only when energy and instructions are available.

Enzymes are biological catalysts. A catalyst is something that facilitates an event that might otherwise occur, but only if you waited for a very long time and got very lucky. A catalyst causes the event to be more probable or to occur more quickly. A community organizer is a catalyst: he or she does not do all the social work but catalyzes other people to do the work. Cats are also catalysts. If you wait long enough, everything in your house will eventually end up on the floor. Cats merely make them end up on the floor more quickly. Cats usually facilitate the law of gravity. Sometimes, as when they carry small objects up to the top of their roosts, they defy gravity (figure 2-1). Enzymes facilitate the natural tendency toward molecular breakdown, but they also defy this natural tendency, as long as the necessary energy and instructions are provided.

Figure 2-1. An example of a catalyst. Catalysts facilitate the movement of objects from high to low positions, and use energy and information to carry objects from low to high positions. Photograph by the author.

The instructions for making these enzymes come from DNA molecules. The DNA that encodes the instructions for one kind of enzyme, or a related set of enzymes, is called a *gene*. Enzymes carry out only the chemical reactions for which the genes provide instructions. In some viruses, and perhaps in all the earliest life-forms, a simpler molecule (RNA) encoded the necessary information. Modern cells still use RNA. The DNA passes the information on to the RNA. It appears that the DNA is a sophisticated control system that was added onto the simpler RNA-based system used by earlier life-forms. Even simpler informational molecules may have preceded RNA.[1]

Information within DNA can be copied. One molecule of DNA can become two, two can become four, and so on. This is what allows cells to reproduce: one cell can become two, two can become four, and so on, each cell receiving a complete set of genes. By the laws of exponential growth, these cellular populations can quickly grow to an enormous size. After just twenty generations, a single cell can become over a million cells. Cells, or organisms, usually cannot reproduce this much because their available habitat and food is limited.

Evolutionary Change Is Inevitable

That is where evolution comes in. When one DNA molecule becomes two, the new strands are almost identical to the old ones. Almost. Occasional mistakes occur during the copying process. These mistakes are called *mutations*. Some of the mistakes are good mistakes, and many are bad. I looked at a poster that listed the major genes on the twenty-three human chromosomes. Most of these genes are named after the disorders that occur when the genes malfunction; for example, for chromosome 17 alone, fifty-three such mutations were listed. The list was very depressing. Mutant genes on chromosome 17 cause Bernard-Soulier syndrome, a type of uncontrolled bleeding; lissencephaly, which means that the brain is smooth, lacking the folds and grooves that allow proper function, and is usually fatal in the first few months of life; one of eleven mutations that can cause Leber congenital amaurosis, a type of blindness; medulloblastoma, which is the most common childhood brain tumor; early onset breast cancer; ovarian cancer; a type of muscular dystrophy; and many other disorders. There are many young organisms of every species that die from mutations. Humans are no exception, even with the advanced medical care now available. Most mutations, however, are small and have little or no significant effect on the organism.

Mutations accumulate in populations of cells or organisms. Before long, you have a population that contains different forms of the genes, or genetic variation. Genetic variation means options. That is, the cells or organisms in this population have options to play the game of survival and reproduction in slightly different ways. The cells or organisms that play the game better will produce more copies of themselves. Cells or organisms with deadly mutations die. Inferior cells or organisms produce fewer copies of themselves than do superior ones. The result is that the superior cells or organisms become more common than the inferior ones. This is *natural selection*: nature selects the superior mutations. That is what evolution is. That is all evolution is. And there is no way in the universe to stop it, so long as no catastrophe wipes out the evolving population.

Natural selection was Charles Darwin's insight. He knew nothing about DNA, but he knew about heredity and about genetic variation in populations of plants and animals. This was enough for him to figure out how natural selection worked.

A population of cells evolves because the superior cells reproduce more than the inferior ones. This makes the whole population better suited to its habitat. For example, near an undersea volcanic spout, the water is very hot, and the bacteria that (by chance) survive best in hot water are superior to the ones that (by chance) survive better in cooler water. Natural selection causes the evolution of bacteria with greater heat tolerance. Natural selection can make a population, therefore a species, better adapted to its habitat.

Speciation Is Inevitable

But there is something else that evolution can do. Suppose that some of the deep-sea bacterial cells wander off into a different habitat, maybe a little farther away from the undersea volcanic spout, and find themselves in cooler water. The mutants that survived better in cooler water are no longer the inferior ones; they are now the superior ones. The cool-water population, if separated from the warm-water population, evolves in a different direction: natural selection causes the evolution of bacteria with less heat tolerance. Natural selection can make two populations, therefore two species, adapted to different habitats.

EVOLUTION CAN OCCUR RAPIDLY

Many people, quite reasonably, wonder why they cannot see evolution happening. The usual scientific answer is that evolution happens so slowly that you cannot see it. While this is true, it is not because evolution has to be slow. Evolution can be very rapid.

Monkeys and Weasels

Creationists never tire of saying that evolution cannot happen, because it would take forever for something to evolve. They like to use the monkey and typewriter (or, nowadays, word processor) example. Suppose you had a room—or a mall, or a continent—full of monkeys typing away at keyboards, randomly striking the keys. How long would it take for one of the monkeys to, by chance, type an entire Shakespeare play? I will not attempt the calculation, but we all know it would be a very, very long time. If the monkey and typewriter scenario were really the way evolution works, then clearly evolution would be impossible. If we had to wait for evolution to, purely by chance, cough up a jellyfish or a sequoia tree or a human, we would be waiting essentially forever.

But, as we saw in the previous section, evolution is not just a matter of chance. The mutations are produced by chance, but natural selection accumulates the good mutations nonrandomly. This changes considerably the monkey and typewriter scenario. If a monkey types a letter at random, but it is the correct letter, natural selection will save it, and do the same with the second letter, the third, and so on. The monkeys, typing at random, may in this way quickly produce a work of Shakespeare, provided that each correct keystroke is preserved. Selection is a ratchet: it saves the good keystrokes and discards the bad ones. In one famous example of this process, evolutionary biologist Richard Dawkins used a computer program that started with a random series of letters. Each generation, the program introduced random mutations into the series of letters and saved whichever mutations caused the string of letters to more closely resemble the Shakespearean phrase "Methinks it is like a weasel." This quote comes from Hamlet, who was looking at the shapes of clouds. The clouds had random shapes, but Hamlet was trying to select the ones that looked like animals. It took only about forty generations for the computer program, now popularly called the "weasel program," to transform a string of random letters into the Shakespearean sentence. Forty generations is not very long. Therefore, in theory, evolution should proceed rapidly. The monkeys could go home before lunch.[2]

Artificial Selection Shows That Evolution Can Occur Rapidly

Evolution can work really fast. Scientists, working in laboratories, have used selection to get populations of bacteria to evolve the ability to use different kinds of sugar as energy sources, or to tolerate different temperatures, in just a few weeks or months.[3] Imagine what could happen on a whole planet over millions of years. What the scientists did is an example of *artificial selection*, rather than natural selection, but the process is the same, only humans rather than nature are doing the selecting.

What is true of bacteria is also true of complex organisms. Starting with pups of wild wolves, humans have used artificial selection to breed all the different kinds of dogs, from Great Danes to Chihuahuas, within just the last ten thousand years. This includes useful breeds of dogs that herd sheep and chase or retrieve prey. The breeds of dogs that are small, shivering, and generally useless are of even more recent origin. The prodigious variety of domesticated plants, as well as livestock animals, has evolved by artificial selection mostly in the last five thousand years.[4] What this shows is that the mutations are out there in the wild populations, and that deliberate, focused selection can make evolution occur very quickly.

Evolution in Hospitals and Down on the Farm

Perhaps the best example of how rapidly natural selection can work is the evolution of resistant organisms, such as antibiotic-resistant bacteria. A mutant bacterium that has the ability to resist penicillin will thrive in the body of a person who is taking penicillin. This is because the penicillin has wiped out the bacteria that cannot resist it. The result is a person who is carrying around a culture of resistant bacteria, which can then spread to other people. This occurs readily when humans are in close contact, such as in hospitals, overcrowded prisons, or schools. This same mutant bacterium, in the absence of antibiotics, is inferior to the nonresistant bacteria (figure 2-2).

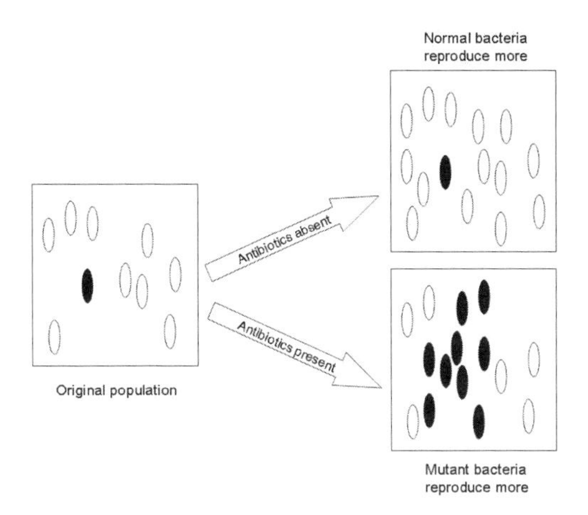

Figure 2-2. Evolution of bacterial resistance to antibiotics. A mutant bacterium has an inferior ability to make cell walls. In the absence of antibiotics (top), this mutant bacterium has an inferior ability to reproduce. However, in the presence of antibiotics (bottom), the mutant bacterium (while still reproducing slowly) survives better than the normal bacteria, most of which die upon exposure to the antibiotics. Diagram by the author.

Humans have developed many kinds of antibiotics, but for each of these antibiotics, there are populations of bacteria somewhere that have evolved resistance to them. Fortunately, not all bacteria have evolved resistance to all antibiotics. But some populations of bacteria, called "superbugs," have evolved resistance to several kinds of antibiotics. Many of you, like me, have known someone who became ill or died from a resistant bacterial infection they acquired in a hospital or nursing home.

It takes only a few years for populations of bacteria to evolve resistance to any particular kind of antibiotic, and it has taken only a few decades for resistance to antibiotics in general to become a major public health problem. Many viruses, such as HIV, have evolved resistance to antiviral medications as well.[5]

In a similar fashion, and almost as rapidly, populations of insects have evolved resistance to pesticides used to control them. There are many populations of insects, many of which spread diseases from one human to another, that will no longer die if sprayed with a pesticide. There are populations of disease-carrying rats that cannot be killed by rat poison. Environmentalist Rachel

Carson pointed out in 1962 that the overuse of pesticides not only polluted the environment but also proved ultimately useless because of evolution: "If Darwin were alive today the insect world would delight and astound him with its impressive verification of his theories of the survival of the fittest. Under the stress of intensive chemical spraying the weaker members of the insect populations are being weeded out. Now, in many areas and among many species only the strong and fit remain to defy our efforts to control them."[6]

In all these ways, from the evolution of antibiotic-resistant bacteria to the evolution of pesticide-resistant vermin, agricultural and medical researchers (some of whom are creationists) have had to take evolution into account when developing strategies to control the spread of infectious diseases and pests. Germs and disease vectors are moving targets, and evolution is the reason for this. Evolution happens in hospitals (and down on the farm, too), not over the course of millions of years but over the course of just a few months. What you don't know about evolution can kill you.

The question, then, is not whether evolution can occur rapidly enough to account for the diversity of life on this planet, but why it has been so slow. There are lots of mutations out there in populations, and selection can operate rapidly. If all the conditions had been just right, life could have evolved into its many complex forms within just a few million years. Why didn't it?

IF EVOLUTION IS INEVITABLE, WHY IS IT SO SLOW?

The answer to the question just posed is that the environmental conditions in which each population lives keep changing back and forth. Sometimes warm, sometimes cool; sometimes wet, sometimes dry. And the interactions of species with one another keep changing. Evolution could make rapid progress if there were some force directing it toward a goal, but instead it wanders around aimlessly—because evolution has no goal.

Darwin's Finches

Consider one of the best-studied examples of evolution: the finches of the Galápagos Islands. In 1835, a British naval ship, HMS *Beagle*, stopped at the

Galápagos Islands, off the coast of Ecuador. The young naturalist onboard, Charles Darwin, had just enough time to explore some of the islands. He took some notes, captured some birds, and rode around on the backs of some of the large tortoises after which the islands were named (figure 2-3).

Figure 2-3. A Galápagos tortoise. Some of the tortoises alive today (many live in captivity at a research station) may have been alive at the time that Charles Darwin visited the islands in 1835, though this is unlikely. Darwin noticed that there were different kinds of tortoises on the different islands in the archipelago. This was one of the observations that got him thinking about evolution and how it might occur. Photograph by the author.

Based on what Darwin saw and what he heard from others, it appeared that each island may have had its own species of birds and tortoises, even though the islands all had a similar climate—brutally hot and dry, except in the misty mountains. Like nearly everybody else, Darwin had been taught to believe that every species had been created by God and placed in its proper habitat. But if that were the case, why would such similar islands have different species? He wondered about it.

When he got back to England, however, Darwin could not wonder about it anymore. He had to figure out an answer. An ornithologist, John Gould, identified many of the birds in Darwin's collection as finches. It became obvious to Darwin that all the finches, which differed greatly in size and in the way they gathered food, were the recent descendants of a kind of finch that lived on the mainland of South America. Evolution had occurred—it had occurred rapidly—and before long Darwin had figured out how it occurred: by means of natural selection. The original immigrant finch populations evolved different ways of getting food. These finches, now called Darwin's finches, have become famous as the group of organisms in which the process of evolution was discovered.

These finches are famous for another reason as well. A husband-and-wife team, American evolutionary biologists Peter and Rosemary Grant, began to study these finches in the 1970s. They were not the first ones to have done so since Darwin; British ornithologist David Lack had stayed for a few months and studied them in the 1940s.[7] But the Grants had something different in mind. They came not just for a brief visit but for a long-term study. They established their base on the island the British called Daphne Major to study the process of natural selection, mostly in the medium ground finch *Geospiza fortis*. In order to study natural selection, it is necessary to demonstrate that the offspring of some of the birds survived better and laid more eggs than the offspring of other birds. Sounds easy? Not on your life. In order to do this, the Grants had to keep track of each bird. They had to know which birds mated with which other birds, how many eggs they laid, and then keep track of their offspring. And then do it again and again, every year. Even on that small island there are hundreds, sometimes thousands, of medium ground finches. The Grants, with the help of colored bands they placed on the birds' feet, became personally acquainted with hundreds of birds each generation, over the course of more than thirty generations.

In order to study natural selection, you have to choose a trait or feature that evolves and that can be measured. The Grants chose to measure beak size. Why beak size? Because a small difference in beak size, even as little as half a millimeter, can make all the difference in what kinds of seeds the birds can crack open and eat. With a slightly larger beak, a finch can crack open and eat the seeds of puncture vines; with a slightly smaller beak, a finch can eat smaller seeds from other plant species.

The birds with the larger beaks preferred to crack and eat the larger seeds: they could get more food with each bite that way. Their beaks, however, were a little too unwieldy to efficiently handle the smaller seeds. The birds with smaller beaks simply could not eat the larger seeds but were fairly adroit with handling the smaller ones. The population of finches always contained a mixture of small-beaked and large-beaked birds—in fact, a whole range of beak sizes.

Enter El Niño and La Niña. El Niño is a set of conditions in which strong ocean currents from the warm waters of the west bring lots of rain to the Galápagos Islands, as well as to much of North and South America. As a result, plants that produce small seeds grow profusely over the ground. El Niño is not necessarily a time of plenty for all organisms. During La Niña periods (when El Niño conditions are not occurring), strong ocean currents bring up cold water from the depths along the coast of South America, feeding marine life such as shrimp and fish and offering a bonanza to birds that eat them. El Niño (and the storms that come with it) is a blessing to some organisms and a catastrophe to others.

The Grants measured finch beaks during a La Niña drought that began in 1977. Small-seeded plants became rare, and large-seeded puncture vines became common. Because more large seeds were available, it would make sense that natural selection would favor large beaks at that time. The Grants did not simply find that the average beak size in the finch population increased during the drought. They found that large-beaked parents laid more eggs, and their large-beaked offspring also laid more eggs, each generation. They had observed natural selection in the act of occurring.

Then in 1982 El Niño brought lots of rain. Very quickly, plants with small seeds sprang up all over the island, and puncture vines became rare. This time, natural selection favored the smaller-beaked birds, and it was the larger-beaked birds that went hungry and laid fewer eggs.

The fluctuation in beak sizes—first larger, then smaller—that the Grants observed was very small, on the order of about a millimeter. But this small change is very important to the survival of the birds. A change of a millimeter could occur within just a few years in a population. The Grants had observed rapid evolution: a rapid increase, then a rapid decrease, in beak size. The Grants had seen, and showed the scientific world, how rapidly evolution could occur.[8]

Even more importantly, their work revealed why evolution usually does not occur rapidly even though natural selection does. The direction of natural selection changes often. Every ten years or so, the climate flips back and forth between El Niño and La Niña. The changes in beak size, on the order of a few years, are rapid, but on the order of centuries, the changes largely cancel one another out. The course of evolution wobbles rapidly along a very slow and often unvarying path. If Nature would just make up its mind, the way animal breeders make up their minds, Nature could breed its species a lot faster than it does. If only Nature had a mind and a purpose, but it does not.

Sometimes Natural Selection **Prevents** *Evolution*

There is more than one kind of natural selection. *Directional selection* causes the average population trait to change, for example, to cause the average beak size in the finch population to increase from one generation to the next. When the average beak size decreases, natural selection has taken a different direction. This is one kind of natural selection that makes evolution occur.

Stabilizing selection occurs when the average population trait remains unchanged. Stabilizing selection eliminates both the largest and the smallest beaks, maintaining a medium beak size. This is the kind of selection that prevents evolution from occurring.

Which kind of selection is happening in the case of the finch beaks? It depends on the timescale you use. Over the course of a few years, it is directional selection. Over the course of centuries, it is stabilizing selection. In fact, most of the evolutionary history of most species is one of long-term stabilizing selection. Most species remain relatively unchanged for the thousands or millions of years of their existence. Directional selection may temporarily change them a little bit, but stabilizing selection keeps them from changing very much. Once a species has found a successful adaptation, natural selection generally eliminates the deviations. This pattern has been called *stasis*, or *equilibrium*.

This is one reason that evolution does not occur as rapidly as it could. Directional selection wobbles around, producing long periods of stabilizing selection. Natural selection is inevitable, all the time. But rapid evolutionary change is not inevitable.

Hurry Up and Wait

Both my parents served in World War II; my father as a soldier (luckily, never deployed) and my mother as a dental hygienist. They told me that life in the military consisted of "hurry up and wait." Others have described the military life as long periods of boredom interrupted by brief periods of terror.

And that is what evolution is like. Most of the time, in most species, it does not seem to be occurring—fluctuating directional selection creates long-term stabilizing selection. But sometimes, rapid directional selection occurs. When? Just as you might expect, and as the Grants observed in the finch populations—it occurs when there is a major environmental change. The environment around a population may change, or a few members of the population might migrate to a new location that has different environmental conditions. When conditions change, directional selection gears up to make the population change along with it. On the Galápagos Islands, the conditions changed back and forth. But sometimes environmental conditions change and then do not change back. Or the migrants do not go back home, remaining in permanently altered conditions. Under these circumstances, directional selection has permanent and rapid effects. It may lead to the formation of a new species, and it may do so rapidly. Such bursts of evolution have been called *punctuations.*

Scientists have a name for the overall hurry-up-and-wait pattern of evolution: *punctuated equilibria.* That is, long periods of equilibrium, during which stabilizing selection is the main process, are punctuated by brief periods of rapid directional selection. New species usually originate during these punctuational events. A species comes into existence by a punctuation of directional selection and then stays fairly unchanged until it either becomes extinct or experiences another punctuation. This pattern of evolution was first pointed out by evolutionary biologists Stephen Jay Gould and Niles Eldredge in the 1970s.[9] It is the punctuations that demonstrate just how rapidly evolution can occur.

INEVITABLE COEVOLUTION

For most species, the most important component of the selective environment is other organisms, rather than nonliving components, such as climate. After all, animals all need food, and this means living organisms, whether plants or other animals. The difference between an organism's nonliving environment and its biological environment is that its biological environment can evolve.

Populations of one species evolve in response to populations of another species, and vice versa. This is called *coevolution.*

In many cases, coevolution occurs between general categories of organisms, such as herbivores and plants, or between predators and prey. (*Herbivore* is scientists' jargon for nonhuman vegetarians.) In most cases, the coevolutionary relationship is not specific; for example, predators do not eat just one kind of prey, prey are not threatened by just one species of predator, herbivores do not eat just one kind of plant, and plants are not threatened by just one kind of herbivore. Plant populations evolve in response to several, perhaps many, species of herbivores. In each of these cases, the important resource is food. Therefore, coevolution is as inevitable as is the need for food.

In many other cases, coevolution occurs between just two species. When this happens, the relationship between the two species is overwhelmingly important to them. Such close relationships are called *symbiosis*, from the Greek words for "living together." Sometimes the relationship becomes so close that the individuals of the species even merge together, resulting in *symbiogenesis*, that is, the genesis of new kinds of organisms from symbiosis. This term was introduced by microbiologist Lynn Margulis.[10] So important have symbiosis and symbiogenesis been in the history of life that I have reserved an entire chapter (chapter 4) for their discussion.

There are as many stories of generalized coevolution as there are species, and the most I can hope to do is to give you just a few examples.

Animals Eating Plants

The first chapter of Genesis in the Old Testament is a hymn to an orderly and bountiful cosmos. Surprisingly enough, it begins by describing a universe that existed before God began to create. This universe had two problems: it was chaotic and empty (the old King James English version says "formless and void"). During the six days of creation, according to this hymn, God solved these two problems. On the first three days, he brought order into the heavenly, the fluid, and the earthly realms: making light separate from darkness, air separate from water, and land separate from sea. On the last three days, he filled the heavenly, fluid, and earthly realms: filling the heavens with stars, the water and air with fish and birds, and the Earth with "creeping things." It is obvious to everyone except creationists that this chapter was never, not even when it was first written, intended as a historical description of the history of the cosmos. It described the work of creation as categories, not as a time line. Modern science provides natural explanations for all these things. About five billion years ago, planets formed from the disc of dust and asteroids, and began spinning (day and night); about four billion years ago, the steam condensed and formed oceans separate from the humid air, and continents arose from the sea. Stars began to form about thirteen billion years ago, and some are still forming. And large organisms such as fishes, birds, and creeping things evolved, starting about a half billion years ago. Astronomy, geology, and biology now provide explanations for the processes of creation described poetically in Genesis.

As beautiful as it is, however, the first chapter of Genesis reflects the human-centered viewpoint of the time in which it was written (about 500 BCE), a viewpoint to which most modern people still cling, even if they are not particularly religious. This viewpoint puts humans at the pinnacle of creation, something that most people (including those who think they understand evolution) still do. The writer of Genesis did not praise humans as much as you might think, however: humans were created on the same "day" as all of the creeping things. Humans did not even get their own special day of creation. But Genesis clearly states that humans rule all of the creation, and uses terms that imply that humans have the right to conquer the creation.[11]

One example of this domination is the creation of plants. Plants appear on Day 3, along with the dry land. That is, in this hymn, plants are not considered to be living organisms; they are just a part of the landscape. In a sense this is true: any landscape without plants is quickly ravaged by erosion.[12] But the God of Genesis does not give plants any role of their own other than to be eaten by animals. Behold, he says on Day 6, I have given you every green plant for food.

But plants are organisms. They've got to make a living like everybody else. They have to lift their leaves to the sun and put their roots into the earth in order to grow. And the whole point of growing is to produce seeds (or other reproductive structures)—otherwise they would become extinct. Plants have populations that evolve. It is not in the interests of the plants to just sit there and allow themselves to be eaten by animals. The plants that natural selection favors are those that can *avoid* getting eaten by animals.

As a result, the world is not just a big salad bowl. Nearly every wild plant has several or many characteristics that make it unpalatable to animals. In some cases, plant defenses are clearly visible: spines on cactuses, stinging hairs on nettles. In other cases, the defenses are invisible but even more effective. Most plants have leaves that are at least mildly toxic. Plants produce thousands of different kinds of chemicals that are toxic to animals. Some of them accumulate poisonous mineral ions out of the soil and put them in little sacs (vacuoles) in the leaf cells, where the poisons will not harm the leaves but will harm any animal that chews the leaf, breaking open the sacs of poison. These defenses do not work perfectly. Almost all plants are prepared for the possibility that some of their leaves will be eaten. As soon as they make food by photosynthesis, they transport the food out of the leaves and down into the comparative safety of stems and roots, rather than leaving it up in full view of herbivores. And almost all plants have lots and lots of buds that can produce new stems, each with new leaves, to replace any that are damaged.

Humans have bred crop and garden plants that are palatable, and this has created for most of us the illusion that plants are just waiting to be eaten. They are like Al Capp's cartoon Shmoos, which fall all over themselves in an attempt to get eaten, with stupid grins on their faces. Be warned: wild plants do not have brains, but evolution has produced in them many defenses that appear to those who study them to be very clever. Wild beans, tomatoes, and potatoes are poisonous (figure 2-4). Some tribal peoples subsist on wild cassava. But they have to boil and squeeze the cassava pulp to remove poisons first.

Figure 2-4. Jimsonweed is a poisonous plant in the same plant family as potatoes, tomatoes, and tobacco. Its name comes from "Jamestown-weed," because British colonists in seventeenth-century Virginia had unfortunate encounters with the toxins that fill every part of the plant. Photograph by the author.

Even the parts of plants that are not poisonous are not simple gifts to feed animals. Many species of plants produce soft, colorful, sweet, fragrant fruits. These fruits make themselves very visible to animals and reward the animals for eating them. Such fruits, abundant and hanging low, are in fact part of our image of paradise. (The word *paradise* comes from the Persian word *pairi-daeza*, which means "walled garden.") But the plants have something to gain from this arrangement. The fruits have seeds in them, which the animals swallow when they eat the entire fruit. It is just too much work to pick out the seeds, especially since the pulp clings closely to them. Plants cannot move around and carry their seeds to new locations, nor can the seeds move under their own power. Many plants, therefore, get animals to carry their seeds to new places by enticing the animals to eat them. They pay the animals by feeding them the sweet and nourishing pulp of the fruit. The animals deposit the seeds, along with a pile of fertilizer, far away from the parent tree. Both the plants and the animals benefit. In other cases, plants take advantage of the animals. Seeds with spiky burrs entangle the fur of mammals, which carry the seeds to a new location while they are trying to scrape the burrs from their fur.

Herbivorous animals have little choice but to eat the plants anyway and put up with the inconvenience. Sometimes they are able to avoid the plant defenses, as when very tiny caterpillars eat their way through the inside of a leaf, avoiding the tough exterior and leaving intricate little tunnels in their wake. Many insects, such as aphids, have sucking mouthparts, which can penetrate a plant stem and tap into the plant's sugar tubes. This sap does not contain toxins. Many animals have populations of microbes in their guts (see chapter 4) that help them digest the chewed plant material. This is how cows eat grass. It is not the cow itself but its stomach bacteria that digest the grass.

Some animals have really clever responses to the problem of poisonous food. Milkweed leaves contain poisons that cause heart problems in vertebrates. Monarch butterfly caterpillars eat these leaves, but they are able to sequester the poisons in sacs in their bodies. When a predator eats one of these caterpillars, the sacs spill the poison and the predator becomes ill. The predators learn to avoid monarch caterpillars. Thus monarch caterpillars commandeer the milkweed defenses to use as their own.

The result is an arms race. Plants evolve defenses, and herbivores evolve ways of circumventing the defenses. Plants then evolve new defenses, and herbivores evolve ways of circumventing these as well. For both the plants and the animals, this interaction is expensive. The plants have to use their energy and molecules to make poisons rather than using them to grow new leaves and roots. Animals have to use energy and molecules to digest the plants despite their defenses. In human societies, spending more money on military defense means spending less on growth (such as infrastructure and education). Plants and animals find themselves in the same predicament as humans do.

This is coevolution: natural selection favors plants that are prickly or poisonous, but not so much that they cannot grow, and it favors animals that can tolerate or avoid the poisons, but not so much that they divert too much energy from their own metabolism, movement, and growth.

Once you have learned about the coevolutionary relationship between herbivores and plants, you can never again look at a serene natural landscape the same way. The Lord is my shepherd, said the psalmist, leading sheep into green pastures. The psalmist also said that God prepared a table for him in the presence of his enemies. But the psalmist could not have guessed that, to a certain extent, the green pastures themselves were one of the enemies. There never was a natural Garden of Eden in which all the plants were uncomplicatedly nutritious.

At least, it might seem, this is true of wild plants. Our domesticated crops appear to us to be our unresisting servants. But even this viewpoint may be mistaken. Michael Pollan reverses the viewpoint and tells the story of crop evolution from the crops' viewpoint. Crop plants have been very successful because they have provided us humans with what we need and desire. (He calls this the "botany of desire.")[13] Maize (corn) gives us what we want, and in return we have destroyed millions of acres of prairie to create a big lawn of corn. There is probably no force of nature that could have rewarded a single plant species with such an overwhelming victory over its competitors.

Animals Eating Animals

There never was a Garden of Eden, but there was, perhaps, a Garden of Ediacara.[14] Ediacaran organisms (named after the place in Australia where their fossils were first recognized) were blob-like creatures that lived in the sea about six hundred million years ago. In this innocent garden, there were no predators. As soon as the predators evolved, it seems that the Ediacarans all got eaten, as if they were Al Capp's hapless Shmoos. I will tell you more about them in the next chapter.

It is easy to see what an attraction it is to an animal to eat other animals instead of eating plants. Animal flesh is much more nutritious than leaf tissues. Leaf tissues have a lot of water and fiber, while animal flesh is a highly concentrated source of protein and fat—even more so than seeds, which are rare compared to leaves. One might even say that many herbivores would be carnivores if they could. I would like to share three stories that illustrate this.

First, a colleague of mine told me about a friend of hers who saw a squirrel run out into the middle of the road, where another squirrel had been run over by a car. Oh, the friend said, the live squirrel is saying good-bye to its dead friend. But, as my colleague well knew, the live squirrel was probably eating the dead one. Natural selection has favored squirrels that are really good at finding and eating nuts. They are not very good predators. But if a nice dinner of meat is presented to them, who are they to turn it down? Second, another colleague, an ornithologist, captures birds in mist nets, studies them, and then releases them. One day, while studying his nets out in the field, he found that many of the birds had been decapitated. He discovered that a cow had been walking along the net and biting the heads off the birds. Once again, cows are very poor predators, unless the feast of meat has been offered up to them. Third, yet another colleague of mine lives out in the country and raises chickens, and she told me that it is necessary to use two layers of chicken wire, with a space between them, to protect chicks from being eaten by deer. Nonhuman vegetarians, like most human vegetarians, are tempted by meat.

Different kinds of prey animals have evolved different kinds of defenses against predators. These defenses are overwhelmingly important to them (which is why we call them prey, as if that is their very identity). Some prey animals, such as mice, are small and wary. Their small size allows them to not only hide but also to produce lots of offspring. Many prey animals have camouflage or even coloration to frighten predators (figure 2-5).

Figure 2-5. Prey animals frequently use mimicry to either avoid detection or, in some cases, to frighten away predators. This click beetle has large spots on its outer wings that, when quickly glanced by a predator, look like the eyes of an animal much larger than the beetle itself. Photograph by the author.

Predators are usually swift, intelligent, and have good eyesight. Each of these adaptations allows them to find and catch prey more effectively. It is true that prey would benefit from having these adaptations as well. Swiftness, intelligence, and sharp eyesight would allow prey animals to escape predators. But in most cases, predators are superior in these respects, and natural selection has favored prey that can see only well enough, and are only smart and fast enough, to hide. By spending less of their time and energy on defense against predators, the prey animals that survive can produce more offspring and find more food. Predators generally produce fewer offspring than prey animals do. Sometimes, prey animals are poisonous, and predators evolve the ability to tolerate the poisons.[15]

Prey defenses do not have to be perfect. Some defenses appear to be almost perfect: some mantises look just like sticks or leaves, a disguise good enough to fool even naturalists walking through the woods. But even a little bit of camouflage is better than none at all. I saw a cartoon once in which a lion told a zebra, "You call *that* camouflage?" Black stripes on white (or white on black, I forget which) honestly do not look like the grasses of the African savanna. Except, that is, at nightfall, which is when the lions are most active. Zebras are blatantly obvious in the middle of the day, but that is when the lions are dozing. Predator adaptations need not be perfect, either. As I write, our cat seems unable to tell the difference between my computer mouse with which it is playing and a real one. Even though computer mice have not been part of the evolutionary experience of cats, an extremely intelligent cat should be able to tell that a bright green object without legs is not a mouse. But cats, such as the hundred million feral cats in the United States, are intelligent enough for their own purposes. To have greater intelligence, a bigger brain, would be a waste of resources for them.

Some prey animals have social defenses. They form large herds in which each animal looks out for the safety of the others, to a certain extent. (I discuss this further in chapter 6.) Lions can subdue an individual zebra or wildebeest, but when confronted by a flood of hooves and confusing black-and-white stripes, where to begin?

The Garden of Eden was, by tradition, filled with vegetarian animals—vegetarian tigers and lions. As you can see, such a garden could not have lasted for very long; inevitably, some of the animals would have evolved into predators. There will never be a world in which, as in the vision of the prophet Isaiah, the lion lies down with the lamb. The natural world is not like the *Bambi* movie, with Friend Owl imparting wisdom to little Thumper. In the real world, Friend Owl would be eating Thumper. And, apparently, even Bambi might have been eating some of his fellow critters.

This is coevolution: natural selection favors prey that can escape or hide from predators, or even fight them off, but not so much that they cannot grow, and it favors predators that can catch the prey, but not so much that they divert too much energy away from their own metabolism, movement, and growth.

Coevolution is inevitable. There will always be mutations that make a plant a little more toxic or an animal a little more resistant to toxins, or that make a predator faster or a prey better camouflaged. There is no way to stop the arms race between plant and herbivore or between predator and prey.

INEVITABLE EVOLUTION OF EVERYTHING

The process of natural selection involves these elements: variation, reproduction, and limitation. We usually think of genes in this context: natural selection favors the best genes, which must cooperate with other genes to create the best organisms within populations. But there is nothing about the process of natural selection that requires that the evolvers be organisms. As it turns out, almost everything evolves.

Evolutionary biologist Richard Dawkins pointed out that it is not just genes but any kind of information that evolves.[16] Dawkins called nongenetic pieces of information *memes*, a counterpart to genes. The result is *cultural evolution*, the counterpart to biological evolution. Cultural evolution turns out to strongly affect biological evolution, because, for most complex animals, the most important part of the environment is the culture of its species.

All that is necessary for something to evolve, according to Dawkins, is a faithful but imperfect copying mechanism for instructions and a system that is ready to obey those instructions. DNA and the cell fulfill these requirements. So do computer programs and computers. And so do memes and the human mind.

Memetic Evolution of Ideas

A meme can be as simple as an idea. Somebody thinks of an idea, which is a mutant form of some earlier idea that somebody else had. This is the variation step. In the population of human minds, this new idea contributes to the memetic variation of ideas that is available in the culture. The person then tells the idea to another person or publishes it or sends it. This is the reproduction step. The human mind has a limited capacity for recalling and using ideas. This is the limitation step. The result is that the most successful ideas get propagated in the culture. The less successful ideas remain rare, the arcane knowledge of a few people, or they become extinct. The most successful ideas are not necessarily the best ideas; they may just be the ones people like the most. For every good and true idea, there are probably a dozen bad and stupid ideas that rattle around in people's brains. Many of us try to eradicate the bad ideas from our minds, but with only partial success.

Animal Learning—and Teaching

Humans are not the only animals among whose minds good ideas can spread. Meet Imo, a very intelligent Japanese macaque monkey. Researchers in Japan were feeding their monkeys, including Imo, with sweet potatoes. Imo figured out how to wash the sand off the sweet potatoes before eating them. The idea caught on, and soon everyone who was anybody in macaque society was washing their sweet potatoes. Similarly, chimpanzees use rocks to break open nuts and use sticks to capture termites. Some forgotten chimp invented each of these memes.

Animals have instincts, but they must nearly always learn how to put these instincts to use. Kittens learn to hunt by watching their mother. An abandoned kitten has hunting instincts (exhibited in their complex and rapid play behavior) and may figure out how to hunt, but not very well. The most successful predatory cats are those that learned hunting skills from other cats. The interruption of cultural transmission of memes has turned out to be a problem for the repopulation of endangered species. Endangered animals in zoos have forgotten how to live in the wild.

In most animal species, the youngsters learn their memes by watching the older animals. Imo did not teach the other monkeys how to wash sweet potatoes; they just watched and imitated her. In a few cases, the older animals teach the youngsters. African meerkats, for example, teach their young how to capture and eat scorpions. The adults provide dead or disabled scorpions to the youngest meerkats, but as the young meerkats grow up, the adults provide them with a steadily increasing proportion of intact scorpions upon which to practice.[17] This is one of the few documented examples of active teaching, as opposed to passive imitation, in animal societies (figure 2-6).

Figure 2-6. Meerkats are among the few animal species in which active teaching, as opposed to passive learning, has been observed. Their social structure allows them to survive in harsh desert conditions. Photograph by the author.

Sets of Memes

Usually, memes are parts of complex sets. As a result, memetic selection favors not only the memes that are most useful to the animals that think or learn them but also those that work best as part of a memetic set. Animal communication is an example. Birds have the capacity to produce a certain range of songs, but they acquire their individual repertoires by listening to other birds. The most complex example of animal communication is human language. All humans have an innate capacity for language, but they always learn their particular languages from other humans. The result is that languages evolve.

Evolution is like a branching tree. The main mammalian branch of evolution split into smaller branches, one of which was carnivores, which split into yet smaller branches, one of which was cats, which split into twigs, one of which was the animal now known as a house cat. The branches become separate species when the animals no longer recognize one another as potential mates. Languages have the same pattern. The main Indo-European branch of evolution split into smaller branches, one of which was Latin, which then evolved into smaller branches such as Iberian and French, which evolved into twigs such as when Iberian evolved into Spanish, Portuguese, and Catalan. Mutations accumulate in languages, and memetic selection preserves the variations that are most useful or most appealing to the speakers. After a while, so much divergence has occurred that the speakers may be mutually unintelligible, as when a speaker of Spanish tries to understand French, or when a Parisian pretends he cannot understand a Frenchman from Dijon. An awareness of the branching evolutionary pattern of languages grew during the nineteenth century at the same time as the awareness of biological evolution. A successful word-meme is one that functions best as part of a language, not necessarily by itself.

Memes Are Everywhere

Humans exist in a sea of memes. For example, music is memes. Somebody invents a new piece of music and performs it. A successful set of music memes gets performed by others. Because of memetic selection, there are more copies of Antonín Dvořák's *New World Symphony* on the market than anything written by Johann Ditters von Dittersdorf. Most folk songs have endured centuries of memetic selection. Memetic selection appears to have condemned the deliberately atonal academic music of the twentieth century, such as the works of Anton Webern and Witold Lutosławski, to long-term obscurity. (You've never heard of them? That is my point.)

To a certain extent, memetic evolution is arbitrary. But memes, like genes, must correspond to underlying reality. In the case of music, harmonic consonance creates more of a feeling of peace and enjoyment than harmonic dissonance. Consonance results when the vibrations of different notes match one another's overtone series. Humans desire more than just peace and enjoyment from their music, however, and a certain amount of dissonance is necessary to create a flow of musical events; resolution of a dissonance is part of the story of a piece of music. You cannot change the physical basis of consonance and dissonance any more than you can change the law of gravity. Rhythm is also based on the natural world and may be even more primal than harmony. The powerful drumbeats of Japanese taiko and Lakota wacipi music bring back fetal memories of your mother's heartbeat.[18]

Computer programs are memes. Computer programmers commonly use evolutionary algorithms that consciously imitate natural selection. The program begins with a simple set of instructions, then generates slight variations, tries them out, then retains the ones that work the best. The weasel program is a simple example of an evolutionary algorithm. The computer iterates this process over and over, producing complex results that were designed not by the programmer but by memetic evolution within the computer. Old movies with battle scenes required (and boasted of) "a cast of thousands," but modern movies often have such large-scale scenes generated by computer programs that use evolutionary algorithms.

Literature is memes. The marketplace is an arena of memetic evolution, as people choose some products over others. Science is memes, as scientists try out hypotheses, retaining and propagating the ones that successfully explain the natural world. Everywhere you look, there are memes, memes, and more memes, all of them evolving.

WHY EVOLUTION LOOKS LIKE IT HAS AN OVERALL DIRECTION

When directional selection occurs, evolution has a direction. It rapidly chooses the winners. But directional selection usually does not last very long. One would not expect evolution to take a consistent, overall direction over the course of millions of years. What it takes to be a winner at one place and time is not what it takes at another.

Many Ways to Be a Winner

And this is just what has happened. Life has not evolved in a consistent direction but instead has diversified. The original bacteria in the primordial oceans evolved into species that lived in very different ways: some lived down in the rocks, some lived in hot water, and some used sunlight to make food by photosynthesis. All of these ways of living were successful. Once more complicated organisms evolved, evolution again took lots of different directions (chapter 3). Some organisms evolved into plants that "eat" sunlight and molecules, while others evolved into animals that eat other organisms. Some of them evolved into searching predators; others evolved into hiding prey. Some of them stayed in the oceans; some of them evolved adaptations to survive on dry land. Some of them evolved into larger organisms, some into smaller ones. Natural selection made some of them smart and some of them stupid.

Evolution has not been a ladder leading upward in a single direction. It has been more like a bush observed from above whose branches have grown in all different directions. The "Tree of Life," the branching pattern of evolutionary changes, is often depicted today as a circle, with "LUCA" at the center. The abbreviation LUCA stands for the "last universal common ancestor." It was not the first life-form, but it was the most ancient life-form from which all of today's organisms evolved. Earlier life-forms have no descendants except those that evolved from LUCA.

In the Image of God

And yet, in the popular imagination, the idea persists that humans are the "most evolved" species and that evolution pushes "upward" toward a human-like state. Worms are inferior to us because they did not evolve as much. Chimpanzees are inferior to us also, but not as inferior as worms. Ask somebody about evolution, and they will probably either say it is a plot of the devil or else they will tell you a story that puts humans up at the top of the stairs. Literally. When the publisher was designing the cover to the paperback version of my first *Encyclopedia of Evolution*, the artist depicted stairs, with monkeys at the bottom and humans at the top. One cannot blame the artist, who merely drew what he or she had been taught to believe. The publisher, not the author, decides the cover images of a book, but this publisher immediately changed the image when I told them that it was scientifically wrong.

This up-the-stairs version of evolution is the one that harmonizes most easily with old creationist views. In the Middle Ages, the natural world was depicted as a *scala naturae*, or "ladder of nature." Rocks were at the bottom, simple organisms above the rocks, mammals and birds above the simpler animals, humans above all other animals, angels above humans, God above angels. Not only was there a vast cosmic order, but there were connections among all the components. Sponges were links between rocks and animals, for example. Scholars thought that all the links must exist in God's perfect and complete creation. The eighteenth-century creationist Carl Linnaeus invented our system of biological classification, which includes Latin names (he called humans *Homo sapiens*, "wise men," despite all evidence to the contrary). He was not at all troubled by the idea that there were half-human/half-ape creatures somewhere in the world (he called them *Homo troglodytes*). He was no evolutionist. He simply believed that God's work was not complete unless God filled in all the gaps.

To many early evolutionary scientists, and to many people today, the *scala naturae* still exists, only God and the angels have been lopped off the top, and evolution has been substituted for creation. This is essentially the pre-Darwinian version of evolution that was proposed by French biologist (inventor of the word *biology*) Jean-Baptiste Lamarck. Simple organisms were continually forming themselves out of the mud, then modifying themselves into ever-more complex forms until they became human, or something like a human. We may speculate that the thing that bothers most creationists about evolution is not the idea that creatures arise out of the mud and evolve toward God, but that most of these creatures seem to be content to stay down in the mud, that evolution has no overall upward momentum. That there is no image of God.

Scientists today understand that there is no single, upward staircase of evolutionary progress. All species are equally evolved. Worms are very good at being worms. They can make a living in ways that humans cannot. Chimpanzees are very good at being chimpanzees. In some ways, bacteria can be seen as the most successful life-form on this planet. There are more bacteria, and more kinds of bacteria, than any other group of organisms. They live in a greater range of conditions and have persisted for three and a half billion years. The human species, in contrast, can live in only a narrow range of conditions and has done so for only about a hundred thousand years, about 0.003 percent as long as bacteria have existed.[19]

And Yet, a Story Is Unfolding

There is, however, no denying that life has *seemed* to make progress. Three and a half billion years ago, only bacteria (and probably viruses as well) existed. By a billion years ago, there were complex cells as well. Five hundred million years ago, almost all life-forms lived in the oceans. By four hundred million years ago, some life-forms lived on land. Bigger and more complex organisms have evolved over time. An evolutionary story is unfolding. If this isn't progress, what is? In this section I will explain how a directionless process of evolution can produce the appearance of progress.

The Accumulation of Successes

The reason evolution seems to lead to progress is that evolution accumulates successes. Bacteria were successful; so were the more complex cells that evolved from them. Natural selection kept both of them. Plants have been successful. So have animals. Natural selection has kept both of them. The Tree of Life gets bigger and bigger. Also, natural selection capitalizes on whatever works, whether it is greater complexity or greater simplicity, whether it is life on land or life in the sea. Natural selection capitalizes on opportunity. Each step toward greater complexity opens up new possibilities. Over evolutionary time, successes and complexities have accumulated. Some of the stories have vanished, such as the saga of the dinosaurs. But most of them are still here, to astonish us today as we behold and study them.

Many of the successes were not simply due to increases in structural complexity (chapter 3); they were due to new types of interaction among organisms. One of the principal sources of evolutionary success has been organisms working together for mutual benefit (symbiosis), even to the extent that they merge together (symbiogenesis) (chapter 4). Sexual reproduction has taken evolution in astonishing new directions (chapter 5). Whole new opportunities for adaptation have arisen as organisms (primarily animals) have evolved ways of cooperating with other members of their own species, something known as altruism (chapter 6).

Each of these processes has resulted in a more complex set of interactions that has made the network of life—Gaia—more resilient in response to the inevitable catastrophes that have fallen from the sky, burbled up from the depths, or resulted from the climatic changes produced by the movements of continents and the rise and fall of mountain chains. Evolution has no guiding mind or innate direction. It has simply made use of new opportunities: complexity, symbiosis, sex, and altruism.

In Gaia's middle age, a new species appeared, one that today fancies itself to be in the image of God, the one for whom everything else exists, the one for whom everything else was created. While much of human superiority over other species is a vain illusion, there is a certain amount of truth to it. Humans are, in fact, the most complex species in many ways. Humans have also been the most extreme examples of what symbiosis, sex, and altruism can accomplish. But evolution has also given humans two abilities that no other species has enjoyed: religion (chapter 7) and science (chapter 8). These adaptations have made *Homo sapiens* at once the most explosive evolutionary success, but they have also made us Gaia's problem child, the one that causes her incalculable stress.

Chapter 3

INNOVATION

Evolution is the story of accumulated successes. Through the long evolutionary ages, more and more complex forms of life have evolved. Each new advance in complexity has opened new realms of opportunity. As explained in the previous chapter, this is why evolution appears to have a direction: there are more ways to become more complex than to become simpler. The simpler forms have persisted alongside the more complex; therefore, natural selection favored complexity not because it was better but because it was new. Innovations meant new opportunities for the species that had them.

These new opportunities can sometimes take surprising directions. We often ask, regarding an adaptation, "What is it for?" But one of the most beautiful things about an adaptation is that once it has come into existence, it can be used for something very different than its original function.

In this chapter, we will survey the history of life on Planet Earth, and the life of Planet Earth, and see how biological innovations have allowed life to become ever-more abundant and complex.

THE EARLIEST LIFE-FORMS

Almost as soon as the Earth had oceans, there were simple cells, similar to the ones we call bacteria and archaea. Whether they quickly evolved or had survived from an earlier era is not known. As simple as bacteria and archaea might seem to us, each of the cells has thousands of chemical reactions. They must represent the end product of a stepwise increase in complexity, which started as much simpler bags of chemical reactions.

A World of Bacteria

Three and a half billion years ago, Earth's surface consisted of air, oceans, rocks, and bacteria. If you went back in a time machine and got out, you would have fried to a crisp a few minutes after you suffocated. There were photosynthetic bacteria releasing oxygen gas into the oceans and into the atmosphere, but they had not yet produced very much oxygen. Since there was very little oxygen, there was no ozone layer, and the full intensity of ultraviolet radiation may have beat down on you from the sun. But suppose you had a gas mask and protective clothing. If you had looked around, you would not have seen much of anything. Some of the bacteria may have been abundant enough to form scum. In some locations, layers of photosynthetic bacteria sandwiched between layers of mineral formed pillow-like mounds known today as stromatolites. They would have looked like puffy green rocks.

In this bacterial world, there may have been lots of kinds of bacteria. Some of them may have lived by decomposing dead bacteria or by killing live ones. Some of them, as I mentioned, carried out photosynthesis. There may have been then, as today, lots of kinds of chemical reactions that provided food energy to the bacteria. But bacteria are small and simple (at least, compared to other forms of life), and there are only so many kinds of things that bacteria can do. This bacterial world persisted, gradually building up an oxygenated atmosphere and making carbon dioxide into organic materials, for a couple of billion years. Infant Gaia had a bacterial body.

Complex Cells Burst Upon the Scene

Then, more complex cells appeared. As I will explain in the next chapter, this event may very well have occurred overnight as a merger of two cells, followed by millions of years of evolutionary modification. Eventually, the DNA of these complex cells was enclosed in a nucleus. This may not seem like much of a change, but it was actually very significant. A nucleus has DNA organized into chromosomes, whereas the DNA of bacteria consists of a single loop. If a complex cell tried to store all its DNA information in a single loop, the result would probably be hopeless confusion. But chromosomes can store DNA information in a more orderly way. ("Orderly" is a charitable description. As any biologist will tell you, there is no way to predict which piece of DNA information will be on which chromosome, and the pieces are constantly reshuffled. Still, the level of organization offered by chromosomes is better than what is available to bacteria.) Cells that have more DNA can make more structures and do more things.

Though complex cells may have gotten started overnight, it took a long time for them to diversify and become common. Some of these cells could eat one another. Complex single-celled organisms can swim around in pursuit of or in flight from one another. They can mate with one another. Bacteria can swap chunks of DNA, but complex cells can merge entire sets of DNA and then divide them back up. All these new combinations of DNA provided new opportunities for mutations and evolutionary innovation. Still, each cell was pretty much on its own.

Single cells can do more than they sometimes receive credit for. Among the earliest evidences of life, in the fossil record, are little tracks left in what was once the mud of shallow ocean bottoms at the edges of continents. Scientists used to think that only wormlike animals could have left these tracks. They had not found any worm fossils, but they assumed the tracks themselves were sufficient evidence. However, it turns out that some single-celled organisms can make tracks. The sea-grape, *Gromia sphaerica*, is a large single cell that rolls along the sea floor like a marble. Such a single-celled organism may have made the tracks.[1]

MULTICELLULAR ORGANISMS IN THE WATERS

Eventually, by about a billion years ago, some cells had begun to clump together. At first, it was a pretty simple thing. Cells divided and became new cells, and instead of swimming off in their own directions, they remained attached together. Such a group is known as a colony. Many single-celled organisms form colonies. But the colonial growth form allowed a new step of evolutionary innovation. The different cells in the colony could start specializing on different jobs. Some of them could specialize on producing or digesting food, some could specialize on swimming, some could specialize on producing reproductive cells. A cell within a colony that specializes on one function can generally do it better than can a cell that tries to do everything.

The First Jaws

In the previous chapter, I introduced the Ediacaran organisms. Living in shallow oceans all over the world about six hundred million years ago, some of these organisms resembled blobs, others resembled feathers, and others looked like onion blossoms at a restaurant. Little is known about them. Did they eat chunks of food? They do not appear to have had mouths. Did they move along the sea floor like animals? They do not appear to have had structures for swimming or walking. They do not look like they ever had to eat anything or to run away from anything. Nobody is even sure whether they were animals in the usual sense of the word. They apparently just sat in the oceans and metabolized as the gentle waves caressed them. The innocent Garden of Ediacara existed before there were any predators.[2]

Then, in an evolutionary instant about 540 million years ago, there was an explosive diversification of multicellular animals. It is called the Cambrian explosion because it was the event that defined the beginning of the Cambrian period of the history of life (table 3-1).

Most of the major groups of marine animals, except fishes, appeared at that time, including sponges, echinoderms (such as sea urchins), and, most notably, arthropods. Arthropods are animals with external skeletons and jointed legs. Today we think of lobsters, crabs, spiders, and insects, but the arthropods of the early Cambrian period were quite different from anything you might see today.[3]

Jointed legs gave arthropods an unprecedented ability to move. Their legs were much more versatile and sturdy than the tentacles of any worm or jellyfish. But some of the arthropod legs served purposes other than movement. Mutations in the genes that controlled development allowed some of the legs to evolve into sensory structures such as antennae. And some of the legs evolved into mouthparts that could capture and manipulate food—in other words, jaws.

As soon as arthropods with jaws evolved, they were able to live as predators. They found a world of helpless Ediacaran blobs just waiting to be eaten. Perhaps this is why the Ediacaran organisms became extinct. Within a brief million years or so, the world looked very different. The shallow oceans were filled with predators and prey. The largest predator was an arthropod that had a big round mouth with sharp teeth on the inner ring.[4] Other arthropods, as well as mollusks, evolved into well-defended prey. Evolution modified their external scales or shells into formidable defenses such as razor-like scales and spikes. Some of them, the trilobites, scurried on the ocean floor like cockroaches (to which they

Table 3-1.

Geological Time Scale with Selected Periods and Epochs (to nearest 5 million years).

Eons	**Eras**	**Periods**	**Epochs**	**Duration (million years ago)**
Hadean				4500–3800
Archaean				3800–2500
Proterozoic				2500–540
		Ediacaran		600–540
Phanerozoic				540–
	Paleozoic			540–250
		Cambrian		540–510
		Ordovician		510–440
		Silurian		440–410
		Devonian		410–360
		Carboniferous		360–290
		Permian		290–250
	Mesozoic			250–65
		Triassic		250–210
		Jurassic		210–140
		Cretaceous		140–65
	Cenozoic			65–present
		Tertiary		65–2
			Paleocene	65–55
			Eocene	55–35
			Oligocene	35–25
			Miocene	25–5
			Pliocene	5–2
		Quaternary		2–present
			Pleistocene	2–0.01
			Holocene	0.01–present

Note: The complete list of periods is presented only for the Phanerozoic Eon.
Note: The complete list of epochs is presented only for the Cenozoic Era.
Note: The Holocene Epoch is also called the Recent Epoch.
Note: The Cenozoic Era is now frequently divided into the Paleogene Period (Paleocene, Eocene, Oligocene Epochs) and the Neogene Period (Miocene, Pliocene, Pleistocene, and Holocene Epochs).

Table 3-1. A timetable of events in the life of the Earth. Table by the author.

were not related) and were undoubtedly good at hiding in crevices. Both predator and prey arthropods had eyes (one species had five of them), to find prey and to see the approach of predators. Welcome to the modern world.

Within another fifty million years, most of the major groups of animals had evolved. There were even vertebrates—the first fishes. (The plural of *fish* is *fish* when you are talking about one species of fish but *fishes* if there is more than one species. A bass, a trout, and a crappie are three fishes.) Even though the fishes had teeth, they did not yet have jaws. They were similar to the lampreys, which are modern fishes with sucker-like mouths. Before too long, fishes had evolved jaws and were caught up in the predator-prey game as well. Some of the fishes evolved to be very large and covered with armored plates.

The First Bones

Many kinds of fishes, such as sharks, evolved in the oceans. Meanwhile, many arthropods and fishes adapted to living in freshwater streams and lakes on the continents.

Sharks have skeletons made of cartilage, not bone. Bone is cartilage strengthened by calcium phosphate. The bony fishes, the group to which most modern fishes belong, evolved in fresh water and later moved into the oceans.[5] This suggests that bones first evolved not to provide a strong structure to the body—sharks are plenty strong—but for some other reason. We have a clue about what that might have been by considering what vertebrates use bones for today. Bones are a way of storing calcium, which can be scarce in fresh water or in an animal's diet. If your heart muscles need calcium, they rob it from your bones, which become weak as a result. Bone density loss is a nearly inevitable part of aging. The evolutionary diversification of bony fishes might have been made possible because bones allowed them nutritional flexibility, not just flexibility of motion.

If you had visited Earth 450 million years ago, during the Ordovician period that followed the Cambrian, the continents probably would have appeared barren, but the oceans would have been filled with fishes and arthropods and seaweeds. Although the species were different from those found there today, the oceans would have looked familiar to you, with the exception that there were no aquatic mammals such as whales and dolphins. The oceans and fresh waters were filled with organisms, creating a network of life, of photosynthesis and respiration. Living waters surrounded barren continents.

MULTICELLULAR ORGANISMS ON LAND

Life was comfortable in the oceans. It was a dreamland of equable temperatures. Organisms were surrounded by water, which supported and fed them. In contrast, dry land was a place of extremes: it could get very hot, very cold, and very dry. But it was a land of opportunity. For plants, it was an environment of bright sunlight for photosynthesis. For animals, it was a place where predators could not follow. For both, it was an escape from competition with others of their kind. But to live on land, organisms needed vastly greater amounts of complexity. Life on land needed to wait for evolutionary innovations.

Support Structures

In the oceans and fresh waters, the buoyancy of the water was enough to counteract the tug of gravity. But air provided almost no buoyancy, and to keep from crushing under their own weight, organisms needed to have support structures. Aquatic animals already had primitive forms of the structures that they needed. Many vertebrates already had bones, and arthropods had external skeletons, which were strong enough to allow structural integrity. External skeletons are not strong enough to support the weight of large animals, and to this day terrestrial arthropods are all small. The only large arthropods were and are in the oceans. But the terrestrial world was open to the explosive evolution of small arthropods.

Plants, however, needed a new kind of support structure if they were to survive on land. Until about 450 million years ago, the most complex multicellular photosynthetic organisms were algae (which are actually protists, not plants). By about 420 million years ago, during the Silurian period, the first terrestrial plants had evolved. They were found only in wetlands and places with moist soil—the hillsides remained barren—and were only a couple of feet tall.

The evolutionary innovation that allowed the first land plants to grow up into the air was xylem. Xylem is the main type of cell found in wood. Xylem cells are hollow pipes with strong walls. Xylem allows plants to grow tall, for two reasons. First, plant cells (for example, the cells in leaves) need water. If plants had to rely on simply soaking up the water from the ground, not enough water could diffuse to the tips of the plants unless they were very short. To reach the top of a plant more than a couple of inches tall, the water has to flow through the xylem plumbing. Second, the strong walls of the xylem hold the plant's stem up. Herbaceous plants, with soft, flexible stems, have much less xylem than do woody twigs, branches, and trunks.

Therefore, aquatic vertebrates already had bones, and aquatic arthropods already had shells, but plants needed to evolve xylem before they could live above the surface of the water. Once this occurred, the wetlands and moist soils of the world filled with the ancient relatives of modern clubmosses, horsetails, and ferns, among whose stems lived arthropods such as spiders. There were not yet any terrestrial vertebrates. By 420 million years ago, thousands of square miles of continental area became green because of the evolutionary innovation known as wood.

There appeared to be one anomaly in the Silurian landscape. All the plants appeared to be small, except one that was apparently ten feet tall. We would assume that such a plant must have had a lot of xylem, as much as a modern tree. But its fossils contained no evidence of wood grain. Some scientists proposed that this organism was not a plant at all but a giant fungus. Tall, lonely white mushroom sentinels stood guard over extensive marshes. But how could mushrooms grow so tall without xylem or some similar reinforcement? No mushroom grows this tall today. The controversy continues, with the most recent view being that this organism did not stand up at all but was actually a rolled mat of smaller plants.[6]

During the Devonian period, 360 million years ago, the wetlands of the world contained tree-sized versions of the clubmoss, horsetail, and fern plants that evolved during the Silurian period.

Preventing Water Loss

All cells have to remain wet in order to function. Some cells can tolerate periods of dehydration, but they cannot function during that time. The outside surfaces of land plants are covered with a cuticle of wax, which forms a protective barrier that keeps the cells wet. A wax cuticle would hold in the water, but it would block the inward movement of carbon dioxide for photosynthesis. Plants require a flow of water, and a completely impervious cuticle would prevent the water from flowing from the soil, through the roots and stem, into the leaves, and then evaporating into the air. It is the evaporation of water from plants that keeps them from overheating in the sunshine. Therefore plants need to have just the right amount of evaporation—not too much, not too little. Their evolutionary innovation was that the epidermis cells, with their waxy cuticles, contained stomata, which are little pores that open and close. When open, the stomata allow water to evaporate into the air; when closed, the stomata allow the plant to be imperviously sealed from evaporation.

Animal cells must also remain wet. Unlike plant cells, animal cells require water with a certain amount of salt in it. When animals began to live on land, they had to bring the ocean with them. To this day, the balance of salts in your cells is reminiscent of the saltiness of the ocean.[7]

Arthropods already had shells, which can be fairly effective at preventing evaporation. But, like plants, animals have to evaporate a little bit of water so that they do not overheat. A completely impervious barrier would, moreover, cause animals to suffocate. Land animals therefore have to have some way of absorbing oxygen without losing too much water. They do this in one of two ways: through the tracheae or through the lungs. Both of these are internal surface areas that absorb oxygen (and release carbon dioxide), while minimizing the external surface area through which water vapor is lost. Insects have tracheae, which are little air tunnels that penetrate their bodies. Spiders have lungs, which are caves of internal surfaces.

It was not until the Devonian period, over 360 million years ago, that any vertebrates ventured onto the land. To do this, they needed lungs. Most fishes have swim bladders, which hold gases that allow them to have just the right amount of buoyancy. Structurally, a swim bladder is equivalent to a lung. But some fishes have lungs instead of swim bladders. These lungfishes today live in the shallow waters of some tropical rainforests. Lungfishes gulp air, and this allows them to live in murky water that contains too little dissolved oxygen for gills to work properly. Darwin speculated that lungs evolved from swim bladders. It now appears that the opposite is true: the earliest freshwater fishes had lungs, and swim bladders evolved from these lungs. The point is that, before the first vertebrates ever climbed out onto the land, they already had lungs.

Walking on Land

Insects and spiders had already been walking on land for millions of years before the first fish dragged itself out of the water during the Devonian period 360 and 370 million years ago. The first steps on land by an amphibian-like fish, or a fishlike amphibian, would not have been particularly dramatic. The evolution of walking was a gradual process. Some fishes lived in shallow murky water, where they had retreated to escape from larger predatory fishes. They already had lungs, and they already had arm and leg and finger and toe bones inside their fins. With these reinforced fins, they could push their heads (which already had nostrils) out of the water to breathe. This allowed them to live in very shallow water where no large predators could follow them. The transition to life on land required only that the arm and leg and finger and toe bones become arms, legs, fingers, and toes. To this day, amphibian juveniles still live in the water, and amphibian adults do not usually live very far from it.[8]

The amphibians that left the water were not seeking out a new life on land but escaping from some aspects of life in the water. But once they were living on land, they discovered new opportunities. There were a lot of terrestrial arthropods to eat in the thick stands of small trees and undergrowth.

Reproducing on Land

Sexual reproduction requires that sperm swim to eggs and fertilize them. For plants and animals that live underwater, this is no problem: the sperm can just swim through the water. But to reproduce on land, animals must copulate or have some equivalent process. The male directly delivers sperm into the body of the female; otherwise the sperm would dry out and die. As will be further explained in chapter 5, many fishes and other aquatic animals copulate not to protect the sperm but to keep them from getting lost. Moreover, the eggs of fishes and amphibians, laid in the water, do not have shells and would quickly dry out on land.

Most reptiles spend their entire lives, and lay their eggs, on dry land. When the first reptiles evolved, about three hundred million years ago during the Carboniferous period, copulation was the only possibility for sexual reproduction. Reptile eggs (except for the few reptiles that have live birth) have hard shells. In addition, reptiles have tough skins that are much more effective than the soft skin of amphibians at resisting water loss. In general, reptiles, unlike most amphibians, can live not only on land but in dry regions of the land surface. During the Carboniferous, they lived mostly in forests because there were not very many dry regions on Earth, and those regions had no plants, no insects—no food.

Some of the trees grew up to a hundred feet tall in the wet forests of the Carboniferous period. Huge masses of their dead trunks and foliage make up much of today's coal deposits. In addition to the tree-clubmosses and tree-horsetails, there was a new kind of plant in the world, with a new kind of reproduction. It was the conifer.

Modern conifers include pines, spruces, and junipers (figure 3-1). The conifers of three hundred million years ago were different, but their reproduction resembled that of modern conifers. Conifers do not release sperm, as ferns do. They release pollen grains, which contain sperm nuclei. The sperm nuclei are protected inside the tough-coated pollen grains. When the wind carries the pollen grains of a conifer, the sperm nucleus does not dry out the way a sperm would. Pollen, therefore, is an immensely successful way in which sperm can travel from one conifer to another without drying out and dying. Flowering plants evolved later, and, like conifers, their sperm nuclei were protected inside pollen grains. I describe the story of flowers further in chapter 4.

Figure 3-1. Ponderosa pines produce pollen grains in male cones (left) and ovules in female cones (right). The pollen grains contain sperm nuclei and are carried through the air by the wind. The pollen grains land on young female cones and the sperm nuclei fertilize the ovules. The ovules grow into seeds.

Ferns produce eggs that remain inside a protective structure; the sperm swims to the egg. Conifers produce eggs inside ovules, which are not merely protective but which contain a good supply of food. When the pollen grain lands on a female cone, and the sperm nucleus descends through a tube to the ovule, the sperm nucleus fertilizes the egg cell. The egg cell becomes an embryo, and the ovule becomes a seed. When a conifer seed lands on the ground, it has a much better chance of survival than does the fertilized egg of a fern. In particular, a seed can withstand long periods of dry conditions. When water finally becomes available to it, the seed germinates. (Flowering plants also produce seeds.)

Therefore, during the Carboniferous, the forests began to change. The tree-clubmosses and tree-horsetails became less common, and the conifers began to dominate the canopy. Further, trees began to grow on hillsides rather than being confined to wet lowlands. By the Permian period, which ended 250 million years ago, most of Earth's land surface was forested. A large desert in the middle of Pangaea, the world continent, was essentially barren. The plants that today live in deserts and other arid zones, such as grasses and cacti, did not yet exist.

Flying

If you had visited a Carboniferous forest, you would have noticed a strange feeling in your body. And the plants and animals would have seemed familiar, although you would have remarked upon the absence of flowering plants and mammals. But instead of butterflies, bees, ants, or wasps, you would have seen huge dragonflies. Huge, that is, by dragonfly standards; their wingspan was only a little over two feet. Insects were the first animals to fly, evolving the ability to do so at least 150 million years before reptiles and birds. With the evolution of flight, insects could more rapidly disperse to new locations and escape from predators. Flight opened up new opportunities for mating as well.

There are two reasons why such large insects could not fly, or even exist, today. First, tracheae are very efficient at supplying oxygen to muscles in a small insect, but they cannot provide enough oxygen to the muscles of a large insect. Second, even the strongest insect muscles and wings could not support such a dragonfly's weight. Why, then, did such dragonflies exist during the Carboniferous?

The answer explains the strange feeling you would have in your body. The atmosphere at that time (as explained in chapter 2) contained about 35 percent oxygen, unlike the 21 percent the atmosphere contains today. You would be experiencing hyperventilation. The tracheae would in fact supply adequate oxygen to the muscles of a large insect if the air was this rich in oxygen. Oxygen molecules, with molecular weight of 16, are heavier than nitrogen molecules, with molecular weight of 14. This relatively small difference means that a rich oxygen atmosphere would be much denser than today's atmosphere—dense enough that delicate wings could, in fact, support the weight of a huge dragonfly.

By the end of the Permian period, the atmosphere contained only about 15 percent oxygen, little enough that no large insects could survive. In fact, no reptiles larger than a small lizard could survive at higher elevations on the mountains. Not that there would have been any reason for them to be there: most of the vegetation on the surface of Earth had died during the Permian extinction.

After this time of extinction, the surviving animals and plants repopulated the planet. Conifers dominated the new forests, and there were many new kinds of reptiles, including the dinosaurs.

Feathers and Warm Blood

After the Triassic period, during the Jurassic and Cretaceous periods, dinosaurs diversified into many bizarre forms. Some of them deserved their reputation for being huge and stupid. The largest dinosaurs weighed a hundred tons, moved slowly, and ate leaves. Such creature cannot survive for long if there is any interruption in the supply of massive amounts of plant tissue. So they had to keep eating, and their food choices included leaves that were over a hundred and fifty feet above the ground. A hundred-ton dinosaur might have a head no larger than a football, and there was a good reason for this: in order to eat leaves so high above the ground, these dinosaurs needed long necks. A hundred-foot-long neck could simply not raise a heavy head that high. The brain and teeth in such a head had to be small. These dinosaurs were big eating machines with small teeth and small brains.

The carnivorous dinosaurs were smaller and faster. *Tyrannosaurus rex* was the largest carnivorous dinosaur, but still it was only one-twentieth the size of the large herbivores. And there were many species of small, fast dinosaurs—about the size of turkeys. In fact, you could be forgiven if, during a visit to a Jurassic woodland, you mistook one of these dinosaurs for a turkey. Many of them had feathers. Only recently have scientists determined that these feathers had red or orange pigments.[9]

At first this might seem strange, since these dinosaurs probably did not fly, although they may have glided. But the feathers did not evolve as a way of enabling flight. The most primitive known feathers, from the dinosaur *Tianyulong confuciusi*, had a simple structure that was inadequate for flight.[10] They were down feathers and they held in body heat. These little dinosaurs had down jackets and were warm-blooded, so they were swift and efficient predators. And when the rest of the dinosaurs became extinct sixty-five million years ago, some of these feathered dinosaurs did not: their descendants exist today, and we call them birds. The evolutionary invention of feathers, therefore, allowed these dinosaur-birds to be warm-blooded and, later, allowed some of them to fly.

Small feathered dinosaurs were not the only animals that were warm-blooded. Throughout the time that the forests of the Earth were filled with large slow dinosaurs and small fast dinosaurs, they were also the home of small mammals. Mammals had hair, which, like down feathers, held in body heat.

Live Birth and Milk

After the extinction of the dinosaurs, the vertebrates that diversified the most were the birds and the mammals. Within ten million years after the Cretaceous extinction, a diverse set of birds had evolved, and mammals also, including whales and bats.

Both birds and mammals were warm-blooded. But one of the most important adaptations that birds and mammals had, which allowed them to compete successfully with other animals, was a large investment in reproduction. Both birds and mammals produce a relatively small number of offspring and take very good care of them when they are young. Each young bird or mammal is an expensive investment on the part of the parents. Young birds are costly enough that, in many cases, both parents take care of the young.

Female mammals have a particularly important set of adaptations for taking good care of the young. The first adaptation is live birth. In all mammals except the platypus and echidna, the egg hatches inside the mother's womb. The offspring gestate inside the mother's body and do not usually emerge until they are large enough to function at least partly on their own. Some birds, and even some alligators, lay their eggs in piles of rotting vegetation that keep the eggs warm, as if they were inside a mother's body. But clearly there is no safer place for a very young animal to be than inside its mother's body or, in the case of marsupials, the mother's pouch.

Mammal mothers also feed their offspring with milk, which is a very rich concentrate of nutrients. This not only enhances the nutrition of the young and their chances of survival but also allows the young to develop and become larger before they have to survive on the kind of food that adults eat.

These adaptations—the production of a few large offspring by live birth and feeding them milk—have enabled mammals to evolve and live in many different habitats. With the exception of south polar glaciers three hundred million years ago and the deserts of Pangaea 250 million years ago, the Earth seldom had deserts or fields of ice. The climate of large parts of the Earth became cool and dry starting about thirty million years ago, and the ongoing cycle of ice ages began about two million years ago. Mammals and birds had adaptations, especially their devotion to their young, which made them ready to take up residence in these cold conditions. One inevitably thinks of penguins protecting their eggs during the Antarctic winter. Not long after the eggs hatch the fledglings are almost as large as the parents.

The complexities and innovations I have described in this section allowed the network of life to form all over the land, in addition to the oceans.

HUMANS

The human evolutionary lineage has been separate from the chimpanzee lineage for over five million years. For the first three million of those years, during the australopithecine era of human evolution, human brains were no larger than those of chimpanzees. That is, for 60 percent of human evolutionary history, intelligence was not a particularly important factor. The evolutionary innovation that set the human lineage apart from other mammals, from the very beginning, was *bipedalism*. Humans have been walking on two legs for over five million years, much longer than they have had any unusual degree of intelligence.[11]

Walking Upright

Bipedalism is more than just walking on two legs. It requires extensive anatomical changes. Since the head is on top of the body instead of in front of it, the opening for the spinal cord has to be at the bottom of the skull rather than at the back of it. Since the pelvis has to support almost all the weight of the body, rather than just the weight of the back half, it has to be stronger. Moreover, humans have big butts. The gluteus maximus muscles connect the spinal cord and the upper legs. Walking upright requires that these muscles be large. Bipedalism is most efficient when the big toe is in line with the other toes, rather than protruding, thumb-like, as is the case with most apes. Besides such anatomical changes, bipedalism required a psychological change. Walking on all fours feels normal to most mammals, although an occasional monkey will get up and start walking on its hind legs. Walking on two legs feels normal to most humans, although there are some people with mental disorders who prefer walking on their hands and feet with their butts up in the air.

Bipedalism appears to have been the crucial evolutionary innovation that allowed human evolution to get started. However, scientists do not know what advantage bipedalism might have offered. Walking on two legs means that humans cannot run very fast. Even elephants can run faster than humans. Bipedalism, therefore, has a large cost associated with it—the risk of being killed by other animals.

Most scientists assume the advantages of bipedalism have something to do with the fact that we can use our front legs as arms, and our front paws as hands. But what would the first bipedal humans have done with their arms and hands? The obvious answer would be to make tools with them, but this is not correct, since humans did not make stone tools until two million years ago. They may have used sticks as tools, but chimpanzees and gorillas, which are not fully bipedal, can do this. Another suggestion is that the first bipedal humans could carry things with them in their arms. This might be especially important for mothers carrying infants. Chimp and monkey infants cling to their mothers' fur. But human mothers do not have very much fur, except for the hair on their scalps, and, as most mothers know, it can be painful and dangerous for babies to hold onto their hair. Bipedalism might have begun to evolve when the first human mothers said, "Let go of my hair! Okay—I'll carry you!" in whatever yammering proto-language they might have had.

The explanation that made the most sense was that humans began their separate line of evolution right about the time that African forests shrank and savannas spread. The story goes that humans were safe inside forests, but they had to run across the savanna, avoiding the predators that were there, and carrying their food and infants with them.[12]

The only problem with this explanation is that humans were still living in forests over a million years after they began to walk on two legs. This discovery was announced in 2009. Skeletons of *Ardipithecus ramidus* ("Ardi"), 4.2 million years old, revealed that human ancestors walked upright most of the time, but that their big toes were still divergent. This means that much of the time they walked in trees, grasping the branches with their big toes. Moreover, an analysis of the chemical composition of the teeth indicated that the Ardi individuals ate mostly food from forests, not from the savanna. Bipedalism, therefore, was primarily an adaptation for living in forests, not for running from one patch of woods to another.[13]

Whatever the reason for the origin of bipedalism, it eventually provided a tremendous opportunity for human evolution. When our hands were no longer simply feet, and when we had evolved opposable thumbs, we had the kind of dexterity that allowed us to make and use tools.

Stone Tools

Once humans had the ability to use stone tools, they could begin to eat meat. Primitive stone tools were not very helpful for hunting—imagine chasing down a wildebeest with just a sharp rock—but they were very helpful for butchering an animal that was already dead, and in particular for getting nutritious marrow out of the bones. The earliest human meat eaters were no match for lions or even hyenas, and probably had to either wait until the stronger carnivores had left or steal the carcasses.

Human brain size and intelligence began to increase about the same time that our ancestors began to use stone tools. Stone tools were the innovation that allowed this to occur. It takes a lot of calories to feed a big brain, and meat provided those calories. A positive feedback occurred: stone tools allowed humans to have greater intelligence, and intelligence allowed humans to improve on their stone tools.

Fire

Another important innovation in human evolution occurred when humans learned the controlled use of fire. Animals are afraid of fire, and when humans learned to create fires whenever they wanted them and to keep them inside of hearths, they could use fire to keep predators away from their camps and to chase predators away from carcasses. Nobody knows when humans began to use fire. Humans have probably been using hearths for a million years.[14]

Another benefit of the controlled use of fire was cooking. Anthropologist Richard Wrangham points out that cooking either meat or vegetables makes them easier to eat and increases their nutritional value. Once humans began to cook their food, more nutrients were available for the continued evolutionary increase in brain size.[15]

Throughout the history of life on Earth, evolutionary innovations allowed not only increases in complexity but also allowed organisms to live in new places and new ways. The advantage of doing so was not necessarily that the new places or new ways were better—living on land is actually harder than living in the sea, for organisms in general—but because they were new resources and opportunities not yet claimed by other organisms.

Chapter 4

SYMBIOSIS

Just as an infant has all the organs and metabolism that an adult has, ancient bacteria had almost all the kinds of metabolism that we find on Earth today. Since that time, no radically new biochemical process has appeared. It was all there during the age of bacteria—DNA, enzymes, photosynthesis. There is nothing new under the sun. Much of the subsequent biography of Earth consists of cells merging together the processes that they already had, producing complexity from primordially simple parts. By joining together, each of them benefited from the other. They formed partnerships that accomplished things that neither could do alone. This is symbiosis: the close cooperation, and perhaps even the fusion, of more than one species into a cooperative unit. It is symbiosis that transformed Earth from a broth of simple cells into everything it is today.

MERGERS

Several of the symbiotic mergers that enabled the evolution of complex life occurred early in the history of the Earth, back when it would have looked like nothing was happening. The mergers were on the microscopic and cellular level.

Photosynthesis

One of the earliest cooperative units involved photosynthesis. (Photosynthesis, you will recall from chapter 2, is the incorporation of sunlight energy into sugar molecules.) There are several relatively simple kinds of photosynthesis found in bacteria today. Somewhere along the line, billions of years ago, bacteria with two different kinds of photosynthesis fused together and formed a complex and efficient type of photosynthesis.[1] The result was the cyanobacteria, and they were so successful that they became—and still are—the most abundant cells on Earth. The meager food web of the Earth became a vibrant one, based on the sugar produced by cyanobacteria. Chlorophyll, which has a pleasant green color, is the molecule these bacteria use to capture sunlight energy; with this symbiotic breakthrough, Earth became, as it remains, a green planet.

This green planet then underwent perhaps the greatest transformation in its history. These cyanobacteria removed much of the carbon dioxide from the atmosphere and replaced it with oxygen. We take oxygen for granted, as if it is something that all planets have. But, as explained in chapter 1, oxygen is a very reactive molecule—it will steal electrons from almost anything, including iron, which is what makes iron turn to rust. Because oxygen is so reactive, it must be continually renewed; otherwise it would react with everything else and be used up within a few thousand years. As a matter of fact, this is what happened on the young Earth—the oxygen reacted with iron in the oceans until nearly all of the iron turned red. We can see the evidence for this by looking at the sedimentary rocks from a couple of billion years ago—the iron in the older rocks is grayish, but the iron in the younger rocks is red.

An observer who looked at the modern Earth from a million miles away would be able to analyze the Earth's atmosphere and determine that it was full of oxygen. The observer could conclude that some process—life—has to continually renew this oxygen. You could tell from more than a million miles away that Earth is a living planet. Astronomers are analyzing the light from planets that circle around other stars, looking for water—which they have found—and for oxygen, which they have not yet found. If they do find oxygen, you will hear a collective shout from the scientific community, because it will mean that Earth is not the only living planet in our corner of the galaxy.

The formation of an oxygenated atmosphere had yet another effect on Earth. It shielded the planet from ultraviolet radiation, which can damage delicate things such as cells. Ultraviolet radiation fuses oxygen atoms into ozone molecules, which then block ultraviolet radiation. Earth is protected by an ozone layer because it has an oxygenated atmosphere.

The iron on Mars is also red, making it the color of blood and explaining why imaginative humans named it after the god of war. But the atmosphere of Mars contains no oxygen. It is possible that ultraviolet radiation has oxidized the iron atoms in the Martian crust.

Complex Cells

But symbiosis didn't stop with the production of cyanobacteria. All the cells in the oceans a billion years ago were structurally simple, like today's bacteria. Some were bigger than others. Then some of the small cells penetrated into some of the larger ones, or perhaps the large cells engulfed the small ones. Either way, the result was usually that the small cells consumed the big ones from the inside, or the big ones digested the small ones. Then one day—and it might have literally been just once, on a single day—some small cell (of the type we now call archaea) entered into a larger bacterial cell. They formed a partnership. Apparently the smaller cell had a more efficient genetic system, and its descendants gradually, perhaps over millions of years, incorporated the DNA of their host cells. The little archaean cell took care of the genetics, and the big bacterial cell took care of the metabolism. The result was a complex cell: the archaean cell became a nucleus. Modern plant and animal cells are complex and nucleated.

And then it happened again. All the cells in the world, then as now, used sugar as a source of energy for their enzymes. One day—once again, it might have literally been a single day—small bacteria that happened to be very good at metabolizing sugar slipped into a larger cell that already had a nucleus. The result was a partnership, in which the bigger cell provided sugar and the smaller bacterial cell very efficiently metabolized it. The small bacterial cells are still there. Today we call them mitochondria. Most plant and animal cells have mitochondria.

And then it happened again! One day—perhaps on a single day—cyanobacteria slipped into a larger cell that already had mitochondria and a nucleus. The result was a partnership, in which the bigger cell provided the raw materials of photosynthesis, and the cyanobacteria produced sugar. The cyanobacteria are still there. Today we call them chloroplasts. Most plant cells have chloroplasts.

Scientists have long puzzled over what they thought must have been the slow, gradual steps by which complex cellular structures, such as nuclei, mitochondria, and chloroplasts, evolved. Where is the evidence of the gradual, intermediate forms—the missing links? But we now understand that the intermediate forms never existed. One day, a small cell took up residence in a large one, and became, in a single stroke, a primitive nucleus, or primitive mitochondrion, or primitive chloroplast. Of course, the first nucleus, the first mitochondrion, the first chloroplast was not in its modern form. Evolution modified them over the course of billions of years. You may be surprised to learn that the way evolution modified mitochondria and chloroplasts was by simplifying them. The primitive mitochondria and chloroplasts lost much of their genetic material to the nuclei of their hosts. They degenerated, and can no longer survive on their own. But they still have—and still use—a few of their own genes, just enough that we can tell that they were once independently living cells, just like an old man who has only a medal to remind everyone that he was once a soldier.[2]

This is where the complex cells of all plants and animals came from. Some of these cells continue now, as they did billions of years ago, to live independently. We call them protists. Most protists are single-celled organisms, but the cells can be quite complex. As a matter of fact, some of these protists have undergone even more symbiotic fusions, sometimes to an apparently absurd degree. The chloroplasts of normal plant cells are evolutionary descendants of cyanobacteria and are, in effect, cells inside of cells. But the "chloroplasts" of some green protists are the evolutionary descendants of other green protists. That's right: protists inside of protists. That is, they are cells inside of cells inside of cells. These complex "chloroplasts" even have their own tiny degenerated nuclei!

And there's more. The claim that cells have moved into other cells and taken up permanent residence in them seems hard for many people to believe. However, we have to believe it, because it has actually been observed in the laboratory! And not just once!

Here are a couple of examples. Kwang Jeon is a scientist at the University of Tennessee who studies amoebas. In the early 1980s he was well on his way to an obscure but comfortable academic career studying amoebas for the rest of his life. Then one day, a bacterial infection broke out among his amoebas. He probably thought, "There goes all my work!" (Most of us scientists have had similar catastrophic experiences.) To his surprise he found that some of the amoebas survived—with bacteria living inside them. When he used antibiotics to kill the bacteria in the surviving amoebas, he found that the amoebas died also. Not only had bacteria moved into and taken up harmless residence inside some of the amoebas, but these amoebas had come to depend on the bacteria. In the course of a few weeks, this man observed a process very similar to how mitochondria might have originated.[3] Just as amazing, Japanese scientist Noriko Okamoto observed a small green protist cell take up permanent residence inside a larger protist cell.[4]

It almost seems as if symbiotic fusions, just like the ones that produced the first complex cells billions of years ago, are going on all the time all around us, and if we watch closely enough, we will see them.

Big Organisms

For billions of years, the oceans were filled with single-celled organisms. There was a lot of symbiosis going on, and some pretty complex cells resulted. But they were still just single cells. And this is the way it was for most of Earth's history. It was not until a little over five hundred million years ago—about the time the Virginia traveler reached eastern Kansas—that a new kind of cooperation began. But once it did, it resulted in an explosion of evolutionary novelty (chapter 3).

It was the evolution of multicelled organisms. This is not a type of symbiotic cooperation between species, because all the cells in a large organism are genetically identical, or nearly so. But it was a way of cells working together as a unit instead of separately. Its beginning may have been very modest: the evolution of the ability of cells to adhere to one another after they have divided. By itself, this innovation would produce just a clump of cells. Some of the earliest fossils, from a billion years ago, are of cyanobacteria that clumped together to form stromatolites, which are big puffy green mats—as some cyanobacteria still do. Some organisms such as *Volvox*, a sphere of green cells, and placozoans, the simplest animals, are barely more than clusters of cells.

But cells can develop differently in response to chemical signals they receive from other cells. This allows the cells to specialize on certain functions. Some of the cells might specialize on protecting the clump. Others might specialize on processing food for the clump. Others might specialize on moving the clump from one place to another. And once you have different cells in a cluster specializing on different functions, voilà, you have a multicellular organism.

The first multicelled organisms, which appeared about six hundred million years ago, right at the end of the last Snowball Earth event, were not much to look at. As explained in previous chapters, these Ediacaran organisms looked like blobs, plates, bags, or feathers. There is no clear evidence that they moved or even ate. Nor is there evidence of internal organ systems. Some scientists have even raised the possibility that photosynthetic algae were embedded within the tissues of Ediacaran organisms. The Ediacarans did not need to eat because they had symbiotic food factories inside them. If this was the case, then the Ediacaran organisms would represent yet another example of symbiosis. Many modern aquatic animals such as corals and mollusks maintain gardens of symbiotic algae within their flesh, so there is no reason that this could not also have happened six hundred million years ago.

SYMBIOSIS AND THE SPREAD OF LIFE ON LAND

By about five hundred million years ago, there may have been a few simple organisms on the otherwise barren surfaces of the continents. By four hundred million years ago, multicellular plants and animals began to live on land. By about three hundred million years ago, forests covered the wetlands on the continents. By one hundred million years ago, the forests covered the hillsides and were filled with plants that produced flowers and fruits. Symbiosis enabled each of these stages of evolution.

The First Life on Land

While the first multicellular life-forms evolved in the oceans, the continents looked pretty barren, as far as we know. Rocks eroded into sand, silt, and mud. There must have been bacteria and protists in the mud and water of the continents. Some scientists, however, have suggested that if you had looked closely, you might have seen a new kind of life on land: lichens.

Lichens look like wavy or curly crusts of peeling paint on rocks (figure 4-1). They can live in hostile environments such as the primordial barren landscapes of the continents. But lichens are not, strictly speaking, organisms. They are partnerships. Each one is a sandwich of bread-like fungus on the outside, with jelly-like algae on the inside. The fungi protect the algae, and the algae feed the fungi. When it rains, the algae come to life and make food. When it is dry, the algae become dormant, and the fungi protect them from dangerous solar radiation. This is what lichens probably did as far back as six hundred million years ago, and this is what they still do.[5] Today, lichens live not only on rocks but also on tree trunks and branches. The earliest life on land, therefore, was symbiotic.

Figure 4-1. Lichens are symbiotic associations between fungi (on the outside) and algae (on the inside). The fungi protect the algae, which manufacture food. This association allows lichens to live in otherwise inhospitable environments, such as on tree trunks, branches, and rocks. Photograph by the author.

The First Plants in the Soil

The first land plants lived in wetlands and were only a couple of feet tall (chapter 3). They obtained minerals from the soil. The soil, however, was not at first very fertile. Good soil is rich because of dark and fragrant humus, which comes mostly from the decomposition of leaves, stems, and roots. But the soil encountered by the first land plants would not have had any humus in it.

Fossil evidence suggests that these earliest land plants had fungi growing in their roots.[6] Sometimes when fungi grow in plant roots, they are parasitic and kill the plants. But these fungi were symbiotic. Symbiotic fungi called mycorrhizae live in the roots of most plants today and help the plant roots to absorb mineral nutrients, especially phosphorus, from the soil. Even today, most plants cannot live in poor soil without these fungi. A few kinds of plants (mostly legumes and alders) have bacteria that live in their roots. These bacteria do not help plants to absorb mineral nutrients; instead, they actually manufacture nitrogen fertilizer inside the roots. The bacteria absorb nitrogen gas from the air spaces in the soil and convert it to a form of nitrogen fertilizer. These plants, in effect, have little symbiotic fertilizer factories in their roots (chapter 1). Root symbioses are extremely important to life on Earth today, and apparently plant life could not get started on land four hundred million years ago without it.

The First Forests

The first forest trees lived in wetlands. When these trees died, they fell into the muck and piled up. These huge piles of trees eventually became massive deposits of coal. Similar accumulations of single-celled photosynthetic organisms have become oil deposits.

When plants die today, they do not form such deep piles of organic matter, except in acidic peat bogs, which are quite meager in comparison with the coal swamps three hundred million years ago. Today, fungi are able to digest lignin, which is one of the molecules that makes wood strong. Three hundred million years ago, fungi may not have been able to digest lignin, allowing the dead wood to pile up and eventually become coal.[7]

Insects were able to eat the leaves of the trees in the first forests. But apparently vertebrates were not. Today, there are vertebrate herbivores all over the place, from cows to giraffes. Why did vertebrates not take advantage of the vast amount of green food in the ancient forests? What happened that allowed vertebrates to begin eating leaves?

What happened was symbiosis. In many cases, animals that eat leaves and wood cannot digest what they eat! Koalas eat eucalyptus leaves all day but cannot digest them. Cows walk around, moo, and eat grass all day but cannot digest the grass. Colobus monkeys stuff their mouths with leaves that they cannot digest. It is the symbiotic bacteria that live in their stomachs and intestines that digest the plant material for them.

The digestive symbiosis is a good deal for all involved (except the plants): the stomach or intestinal bacteria get a warm, moist, protected chamber to live in, with a continual source of chewed-up food coming in; the animals get nutrients from the food the bacteria digest. The largest dinosaurs, over one hundred million years ago, ate leaves (chapter 3). We do not know, but strongly suspect, that the only reason they could survive on this diet was that they had symbiotic bacteria in their digestive systems. When grasslands became common on the Earth about thirty million years ago, grazing animals spread and diversified—and this could not have happened without their symbiotic stomach and intestinal bacteria.

Flowers and Fruits

Some plants reproduce by making copies of themselves. But most plants, like nearly all animals, depend on sex for reproduction. As explained further in chapter 5, sex allows new combinations of genes to be produced. In animals, sperm swim to eggs. This also happens in some plants such as mosses and ferns. But in seed plants, the sperm nucleus is enclosed in a pollen grain, and the egg nucleus is enclosed inside an immature seed. Sperm must swim under their own power, but pollen grains can be carried from one plant to another (chapter 3).

Wind carries the pollen of most conifers, such as pine trees, but wind pollination is very wasteful. Most of the pollen ends up everywhere except where it is supposed to be. Its target is a young female pinecone, but most of it ends up on the sidewalk or on leaves or in your lungs. But the trees produce so much pollen that even a success rate of one in a million may be enough to fertilize the seeds of the next generation. One summer, I was in the Black Hills of South Dakota, with its millions of Ponderosa pines, during pollination season. I had a blue car. The air turned yellow and the car turned green. Insects pollinated some conifers in ancient forests, but most relied on the wind.[8]

About 120 million years ago, new kinds of plants evolved in the forests. These were the flowering plants. Some flowering plants, such as cottonwood trees and ragweeds, rely on wind pollination, just as do the conifers. But many flowering plants formed a symbiotic relationship with animals for pollination.[9] These flowering plants entice animals, usually insects or birds, to carry their pollen. To do so, they have to attract the animals and get the animals to stick their mouthparts inside the flower. Most flowers, therefore, are brightly colored and have a perfume fragrance. Most flowers also reward the animals with food, or at least with empty sugar calories, in the form of nectar.

But the purpose of the flower is not simply to feed their pollinators. The flower has work for them to do. The flower dusts the animal with pollen. It is nearly impossible for a pollinator animal to reach into the flower and get nectar without pollen getting on some part of its body. There are, however, some bees that manage to avoid the pollen. They chew holes in the side of the flower and lap up the nectar, without coming in contact with the pollen. Aside from these examples, however, pollinators become the unwitting, perhaps unwilling, carriers of pollen.

Flowers also have adaptations that encourage the pollinator to travel to another plant of the same species. One way that flowers accomplish this is to not give them very much nectar. The pollinator cannot sit on or in one flower and have a feast. A few gulps, and it is time to move on. Some flowers, such as wild tobacco plants, even have nicotine in their nectar, which creates a kind of love-hate relationship with the pollinator.[10] The pollinator stays just long enough to quickly drink the nectar, then flies away (we may imagine) in disgust, to another flower. When it reaches the other flower, it gets dusted with more pollen, but also the pollen that it has brought with it rubs against the female parts of the second flower.

Natural selection favors flowers that encourage animals to carry pollen to another flower of the same species. That is, natural selection rewards pollinator fidelity. It does little good for the plant if the pollinator visits all different flower species indiscriminately. And this is where symbiosis comes in. Pollination is not a matter of every pollinator visiting every flower. Most flowers have shapes and colors that attract specific kinds of pollinators—which in turn specialize upon those species of flowers and form a symbiotic relationship with them. Hummingbirds like red, trumpet-shaped flowers, and specialize mostly upon them. What is the benefit for the plants? The pollinator carries most of the pollen to another flower of the same species. What is the benefit for the pollinator? It gets a lot of food for itself from flowers that most other pollinators do not visit (figure 4-2). There are many orchid species that are pollinated by just one kind of bee, and many figs whose flowers are pollinated by just one kind of tiny wasp.

Figure 4-2. Butterflies, such as this monarch, have very long tongues that can probe deeply into and drink nectar from funnel-shaped flowers that short-tongued pollinators cannot enter. Butterflies usually avoid competing with short-tongued pollinators by seeking out funnel-shaped flowers. They do not always do so, as in this photo, in which the monarch drinks nectar from a flower that does not have a long funnel. Photograph by the author.

Plants also need to have their seeds dispersed to a new location. The wind carries not only the pollen but also the seeds of most conifers. In some cases, birds pry seeds from the cones of modern pine trees and plant them in new locations (hoping to find them again and eat them but frequently losing track of them), and this may have been true in the past as well (figure 4-3).

Figure 4-3. Some pinecones produce seeds that are carried to new locations and buried by birds. The cones of these pines, such as *Pinus albicaulis* in the Sierra Nevada, produce sap that prevents squirrels from eating the seeds. Photograph by the author.

Many flowering plants rely on animals (usually mammals) to carry their mature seeds, which are produced inside fruits, to new locations. Some fruits release seeds that have fluff or sails, and the wind carries the seeds away. But in many cases a bird or mammal eats a ripe fruit that contains the seeds. In these cases, the animal does not chew up the seeds. The seeds pass harmlessly through the intestines of the mammals and come out the other end in a new location and with a little pile of fertilizer to get them started. Or the animal carries a nut and buries it, intending to eat it later but loses track of it. Often, only a narrow range of animals disperses each kind of fruit. This is what makes dispersal an example of symbiosis. In some cases, just one species of disperser has evolved in close symbiosis with just one species of fruit. For example, only agoutis open brazil nut capsules and durian fruits are overwhelmingly favored by orangutans.

Symbiosis may be the most important reason that there was an explosion of plant and animal diversity starting about 120 million years ago. There are hundreds of thousands of species of flowering plants, and even more species of pollinating insects. Without doubt, specialized pollination and dispersal symbioses resulted in new species of flowering plants and of insects, birds, and mammals.

Dinosaurs became extinct sixty-five million years ago when an asteroid hit the Earth. But they were already in decline before this happened. Most dinosaurs were too large to act as pollinators, and there is little evidence that they ate fruit. Did they perhaps get left out of the new symbiotic game in town?

Pollination of flowers and dispersal of seeds is not always mutually beneficial. Some flowers trick insects into pollinating them by offering false rewards. Some orchids look—and smell—like female wasps, luring male wasps to mate with them. The male wasps, mating cluelessly with one orchid after another, transport the pollen but receive no reward. There are even some flowers that kill their pollinators. Fly-pollinated flowers produce scents that are repulsive to humans but that lure female flies with the promise of a corpse in which to lay her eggs. In many cases, the fly lays her eggs in the flower. The little maggots hatch and, finding no meat, starve to death.

SYMBIOSIS EVERYWHERE

There are examples of symbiosis wherever you look. I will mention three examples that are interesting, even if they have not (as in the previous examples) transformed the world.

Defense

Many plants have formed defensive symbiotic relationships with animals. For example, many tropical trees (most notably acacias) are covered with ants. The ants live inside the tissues of the trees and consume nectar and protein produced by the trees. In return, the ants attack any herbivore that may begin to eat the tree. All the herbivore has to do is to brush against the leaves, and the ants come swarming out.[11] The leaves of some plants even produce little domiciles for mites (called acarodomatia). The mites attack herbivores that might eat the leaves or fungi that might infect them.[12]

Defensive symbioses can be very important to the trees. Acacia trees live in tropical woodlands that are seasonally dry. Most of the trees in these woodlands shed their leaves during the dry season. Part of the reason they do this is because the leaves would otherwise cause the plant to lose too much water. But another reason is that, during the dry season, a green tree is a tempting target for thirsty herbivores. Acacia trees, however, remain green even during the dry season, protected by their ant armies.

Cowboys

As described in chapter 2, aphids are tiny insects that tap into the pressurized sugary sap of plant stems. An aphid can do this with a slender stylus mouthpart that slips into the plant's sugar tubes. The aphids are incredibly lazy; they do not even have to swallow. They just cling to the stem and let the plant pump the sweet liquid through their guts. The sap goes through the aphids so fast that it is only minimally digested. When present in large numbers, aphids can significantly harm the plant.

The sap is also a good food source for ants, but ants do not have the right kind of mouthparts to slip into the sugar tubes. Instead, the ants obtain sugary sap from the aphids. The sugar that drips from the aphids is nearly unaltered plant sap. In return, the ants defend the aphids. The ants are like cowboys tending herds of aphids. Many kinds of ants, and many kinds of aphids, have struck this symbiotic bargain. It is very common to see this happening in gardens.

You

You, like every multicellular organism, are a symbiotic partnership. You are a coordinated mass of about seven hundred trillion cells. About seventy trillion of those cells are human: your skin, muscles, blood, nerves, and so on. Most of the other 630 trillion are symbiotic bacteria. Only 10 percent of your cells are human. Do not be alarmed. Those bacteria are mostly harmless, and are often beneficial, to your health. Recent research indicates that your body is the home of almost a thousand species of bacteria.

The bacteria specialize on different parts of your body. You are, as it were, a little planet, with different habitats. Your exposed skin, the skin underneath your hairs, the skin in your genital areas, the inside of your mouth, and the inside of your intestines are all different kinds of habitats, with their own chemical, temperature, and moisture conditions, and are inhabited by different sets of bacterial species. Even the inside and the outside of your elbow have different species of bacteria.[13]

Many of the bacteria are just along for the ride: they consume dead skin cells or undigested food. The greatest harm they may do is to give you gas or make you smell bad, but you can control this by the types of food you eat and by periodically showering away their excess populations. A few of the bacteria are potentially harmful—millions of people are carrying around MRSA, the worst kind of antibiotic-resistant staph bacteria, in their noses. But even these bacteria are largely harmless, so long as they stay on your outer body surface and do not get down into your tissues. In fact, the other bacteria keep the populations of the dangerous ones in check. If you were to sterilize yourself and eliminate all the bacteria (which has never been done and is probably impossible), new populations of some of the dangerous bacteria would later explode on the wasteland of your ultra-clean skin and conquer your intestines, and then your tissues, in a gangrenous barrage. (Yes, in fact, gangrene bacteria live in your intestines, but they are harmless if other bacteria keep them under control.)

There is experimental evidence that some of the bacteria on and in the bodies of animals are beneficial. Scientists have raised rats (called gnotobiotic rats) in sterilized cages and fed them sterilized food. The rats were delivered by sterile caesarian section so that they would inherit no bacteria from their mother. These germ-free rats, one might think, should be very healthy. But the opposite was true. They were weak and required special vitamin-fortified rat chow to survive. This is experimental demonstration that some intestinal bacteria produce vitamins in sufficient quantity to support normal rats. And without doubt, the same is true of humans and our symbiotic bacteria.

SYMBIOTIC PLANET

Throughout evolutionary time, symbiosis has sparked diversification and explosive increases in complexity in the web of life on Earth. And to this day, every organism is a standing, walking, or flying corporation of different species symbiotically united together. Earth is not a goddess and does not have a body, but its web of life sometimes—perhaps even most of the time—acts as if it does. Cooperation between species has been the key to most of the diversification of life on this planet.

Biologist Lynn Margulis, cocreator of the Gaia Hypothesis, takes it further. Symbiosis has not only been important in the relationships of some organisms to others; symbiotic relationships have made the network of life, which some scientists call Gaia, possible. Gaia, Margulis has said, is just symbiosis seen from outer space.

Chapter 5

SEX

WHY SEX?

"Why sex?" is a question that does not occur to most people. James Thurber and E. B. White gave a humorous answer in a 1929 book titled *Is Sex Necessary?* As Thurber and White saw it, the main reason for sex was that it makes men feel good sometimes and bad most of the time.[1] But my purpose in asking this question is to gain a biological and evolutionary understanding of sex, especially the surprising forms it has taken in the human species. Where did sex come from, and how has it transformed the web of life on this planet?

Sex and Evolutionary Innovation

In the beginning, there were cells. As we discovered in chapter 4, some of these cells, from different species, merged together to become complex partnerships, from which our modern complex cells are descended. Cooperation of different species, and sometimes the merging of their cells, has been one of life's greatest sources of evolutionary novelty.

But there has been, from the beginning, another source of evolutionary novelty. It is, in fact, a different kind of selection. Natural selection, via symbiosis, has led evolution down the paths of efficiency. But *sexual selection* has led evolution down the paths of outlandish beauty and novelty. Natural selection has made life work; sexual selection has made life interesting. Charles Darwin explained natural selection in his 1859 book, *On the Origin of Species by Means of Natural Selection*. He explained sexual selection in his 1871 book, *The Descent of Man and Selection in Relation to Sex*.[2]

Sexual reproduction occurs when two cells combine their genetic material. The result is not new genes but new combinations of genes. Even bacteria do this sometimes. Some bacteria even grow tubes through which they exchange bits of DNA.

When two complex cells merge, however, the result is a doubling of DNA content. Cells cannot continue merging and increasing their DNA content generation after generation. In order for cell mergers to be sustainable, there also has to be a special sexual kind of cell division that reduces the amount of DNA. Since most cells have two copies of each gene, sexual cell division produces cells with just one copy of each gene. These are the sex cells; in animals, eggs and sperm. Eggs and sperm cells can fuse with one another, producing fertilized egg cells with the right number of genes but in a new combination. These new combinations of genes offer new opportunities for evolution.

Most flowers have both male and female reproductive parts, so it would appear that they could just reproduce by pollinating themselves. But this is not what they do. In most cases, when a flower opens, its male parts (the stamens) and its female parts (the pistils) are not activated at the same time. The stamens may release pollen before the pistils become receptive to it, or the pistils may become receptive before the stamens release pollen. That is, most flowers are both male and female, but not at the same time.

Sex is everywhere. This is puzzling because life would be easier without sex. If every plant produced asexual seeds (as dandelions do), and if every animal was a fertile she-male that just popped out copies of itself, life would be a lot easier. Sexual organisms cannot produce as many offspring as asexual organisms because it takes two, rather than one, organism to produce offspring. For that reason alone, sex reduces potential reproductive output by half. The enhanced genetic diversity that comes from sexual reproduction hardly seems worth it. And yet it must be, for almost every species has sexual reproduction at some point, even if they produce copies of themselves without sex for many generations.

The physical environment—light, temperature, moisture, nutrients—simply does not change fast enough that evolution "needs" sexual reproduction to keep up with it. Mutations and natural selection within nonsexual lineages would probably do just fine. It is the biological environment that puts a premium on rapid evolutionary adjustments. In particular, all kinds of organisms have to keep evolving, rapidly, to remain resistant to parasites. The fungi that attack plants, and the bacteria and tiny arthropods that attack animals, have rapid generation times and can evolve new adaptations very quickly.

They would soon breach whatever protection the host plants or animals may have to them. The plants and animals must therefore evolve new defenses—usually chemical ones, similar to our immune systems—all the time. This is the reason that evolution is never finished. An organism might be perfectly adapted to its physical environment, but it will never be adapted to its parasites. Even if a species appears to not be evolving, it is: it is developing new invisible chemical defenses. It is like the Red Queen in Lewis Carroll's *Through the Looking Glass*, who runs just to stay in place. Scientists have adopted this image (the "Red Queen Hypothesis") to explain why evolution has to keep going and going and going, and therefore why organisms need the great amount of genetic variation that sexual recombination provides.[3]

The exception proves the rule. DNA studies indicate that bdelloid rotifers (a kind of microscopic aquatic animal) have not had sexual reproduction for at least thirty million years. How could they have survived so long without sexual recombination to allow them to evolve faster than their parasites? And they indeed have parasites: under moist conditions, fungi can kill a population of rotifers in a few days. Instead of evolving new defenses against the parasites, the rotifers dehydrate and blow away in the wind, leaving their parasites behind. The dried-out rotifers can then land in a new, wet location and spring back to life—without their parasites. These rotifers have survived without sex because, instead of evolving away from their parasites, they are literally blown away from them.[4]

The Birds and the Bees

Sexual recombination requires that sex cells find one another, and that they have enough of a food supply that the merged cell can begin to grow into a new organism. In most species of organisms, the two kinds of sex cells specialize upon these two different functions. Sperm are very good at traveling. They are numerous, small, and have thrashing tails. Eggs usually remain in one location; they are big because they have a supply of food. Sperm can travel more effectively, and eggs store food more effectively, than would nonspecialized sex cells.

In most animals, the differences between the two genders are genetically based, but that basis is not the same in all species. In some species of animals, the genes that make the difference between male and female are scattered in different locations in the nucleus; they have no sex chromosomes. In some other species of animals, gender is determined by the environmental conditions in which the fetus develops: warm temperatures cause many reptile eggs to develop into males, while cooler temperatures produce females. Many animal species have sex chromosomes. One of these chromosomes is stubby and has few genes (in mammals it is called Y); the other is a full-sized chromosome (X). Female mammals are XX, and male mammals are XY. (Anthropologist Ashley Montagu, in his book *The Natural Superiority of Women*, pointed out that Y is not equal to X.)[5] Female birds have one large and one small sex chromosome, while male birds have two large sex chromosomes, the opposite situation from mammals. And in bees, females have two copies of each chromosome, while males have only one. (As you see from the foregoing, if you want to understand human sex, learning about the birds and the bees is not very useful.)

A SEXUAL HISTORY OF ANIMAL LIFE

Sex didn't look all that interesting when it got started. It was just two cells of roughly equal size fusing together. Then it was two cells of different sizes fusing together. When complex animals filled the oceans, some were male, and they released large numbers of sperm into the water; others were female, releasing a smaller number of eggs into the water.

Then some animals evolved a way of making sure that the sperm got to the eggs by direct delivery rather than by chance. This was the origin of copulation. Many male fishes delivered sperm directly into the body of females, where they fertilized the eggs. The female fishes then laid the fertilized eggs. As explained in chapter 3, once animals began to live on land, copulation became not just an option but a necessity. Sperm will dry up and die unless they are directly delivered to the moist reproductive tract of a female.

Why Males Fight

Copulation also allowed one of the earliest examples of sexual selection to occur: by copulating with the female, the male fish could make sure not only that his *sperm* got to where they were supposed to go, but that *his* sperm, rather than those of another male, got there. At this point, sexual competition began. The bigger male fish could drive away the smaller male fish from his territory. Furthermore, since sperm are cheap to produce, a male fish can fertilize many female fish, each carrying expensive eggs. A dominant male's territory could house numerous females, while the subordinate males were driven out into the periphery where they lived out their lonely days. This drama, males fighting one another for access to females, continued to be played out as amphibians moved onto land, in reptiles, in birds, and in mammals, right up to the male gorilla, which weighs twice as much as the female gorilla.

Sexual selection therefore favored big males that were, using Darwin's inimitable term, pugnacious. A lot of evolutionary novelty has resulted from males competing with one another. The large antlers of deer are an example: they serve no purpose other than for males to butt their heads together or push each other around, over and over, until one of the males relents. The result of such competition may be a harem, where one male has many females and keeps other males away by force. Among elephant seals, such competition can be a bloody mess, as males throw themselves against one another. In this way, sexual selection produced relatively violent males across most of the animal kingdom.

Why Males Do Not Always Fight

Males do not always fight. Their competition can take other forms; for example, sometimes they scramble. Males of migratory animal species, such as many birds, usually arrive on the scene earlier than females. By the time the females arrive, the males have already sorted out their territories.[6]

There is certainly more than one way for male animals to ensure that females are fertilized by their sperm more than by the sperm of other males. Males do not need to be large and combative. Instead, they can go around to lots of different females and fertilize them. In such a case, chances are that another male has already been there, but the new male can do one of several things to ensure his superiority over the other males. Three examples are common in insects. First, the male insect can scoop out the sperm of the previous male with his complex genitalia. Second, for good measure, the male insect often plugs up the female reproductive tract with glue. Third, many male insects remain attached to the females after they have deposited their sperm. You may have seen two butterflies or two dragonflies flying around in the midst of copulation. While this may look like love, the real reason for this behavior is that, because copulation itself does not take very long, the male wants to make sure no other males get a chance to copulate with his female.

The most successful male may simply be the one that deposits more sperm than any previous male. Competition between males to deliver the most sperm can be found in many kinds of animals, including chimpanzees. Male chimps have enormous testicles and penes. Why should they be so large, when all that is necessary is to deliver sperm to the eggs? The reason is that chimps mate promiscuously. A successful male chimp is one who can overwhelm the sperm of another male, with whom a female has just mated. In contrast, the penes of gorillas are quite small, not much larger than those of some male ducks. Male gorillas use force to keep other males away; male chimps use sperm count to overwhelm the sperm of other males.[7]

Humans are somewhere between gorillas and chimpanzees in terms of both anatomy and behavior. Human male genitalia are intermediate in size between those of gorillas and of chimpanzees. Many traditional human cultures have harems; a few are more or less promiscuous. This was probably also true in prehistory.

Female Choice

Evolutionary creativity does not end there. In the previous examples, the males were in charge and the females appeared to accept their fate. But this is not the entire story. Females know that the males want them and will compete to fertilize them. The other half of the equation, therefore, is female choice. Many female animals choose from among their male suitors. Darwin convinced many of his readers about male competition for access to females, but he also wrote about female choice, which his Victorian readers were slow to pick up on.

How does female choice work? In many cases, males have to demonstrate to females that they have something of value to offer other than sperm. Since they produce the eggs and usually take care of the young, female animals want resources and protection. They also want mates that have good genes. Female choice of males that provide resources, protection, and good genes has produced most of the evolutionary novelty and outlandish beauty in the animal kingdom.

One way that males can convince females to choose them is to defend a large territory. This demonstrates to the female that he is healthy and that resources and protection are available to his females. Some insects are territorial. Male dragonflies will patrol a precisely delineated territory. One of my earliest memories of natural history was sitting on the porch and watching a dragonfly go back and forth, back and forth, over our landlocked driveway, which could not possibly have been a very good territory for a species whose females lay their eggs in water. Many vertebrates are also territorial; the examples most noticeable to us are birds. Male birds not only defend their territories but announce them by perching conspicuously and singing. A male with a bigger or better territory fertilizes more females and has more offspring because the females choose him.

Exuberant Beauty

But there is a lot more to the story than this. If it were merely necessary for a male bird to announce his territory in order to be chosen by a female, a simple squawk or two would be sufficient. But few birds, other than crows and starlings, are satisfied with such a meager repertoire. Male birds have a vast array of songs, whose beauty has captivated the human imagination since prehistoric times. Birds have the instinctive ability to produce songs, but they must learn the specific songs they sing. (Sometimes a young bird learns the song of the wrong species and may even end up mating with a female of the wrong species.)[8] Some male birds have a large number of possible songs and invent a few of their own. Perhaps the most conspicuous example of this is the mockingbird and its relatives. Each male mockingbird has dozens of songs that differ strikingly from one another, and many of which mock or imitate the songs of other species. The same is true with the mockingbird's quieter relatives, the brown thrasher and the catbird. Mockingbirds often repeat each song about four times, thrashers twice, and catbirds just once. Listening to a catbird can be a bewilderingly beautiful experience. Every half second, he sings a new song, with gurgling clarity, interspersed with a few mew-like calls. Parrots are notorious for learning to imitate sounds, even human speech, with remarkable fidelity. Why do so many birds produce whole symphonies when a simple shriek would do?

Nor is this the only kind of excess for which male birds are famous. Male birds, especially in breeding season, are noted (again with the exception of crows and starlings) for showy plumage (figure 5-1).

Figure 5-1. This male red-winged blackbird uses three methods to defend his territory and attract mates: bright red epaulets on the shoulders; ruffling his feathers in a way that makes him appear larger; and uttering a loud call. Photograph by Douglas R. Wood.

Every male cardinal bird is red, but some of them are astonishingly crimson. Scissortail flycatchers have double tails that are as long as the rest of their bodies—some quite a bit longer. Male painted buntings are blue, red, and green, and some are more vivid than others. Everybody knows about male peacocks, some of which out-display the others.

The songs and plumage of male birds is excessively beautiful to human ears and eyes but also to those of female birds of their species. This makes intuitive sense, but it has also been demonstrated experimentally. Scientists artificially shortened and lengthened the long tail feathers of some male birds. They found that the male birds with shorter tails mated with fewer females, and those with longer tails mated with more females. Just to prove that their results were not due to the process of cutting and gluing the feathers, they did a control: they cut and pasted tails without altering their length—and this did not alter the reproductive success of the male birds.[9]

What could possibly be so valuable about the ability of male birds to sing beautifully and produce showy feathers? Sometimes their feathers are so excessive that they put themselves at risk. A brightly colored male (or a loudly singing one) cannot easily hide from predators, and in the case of the birds-of-paradise of New Guinea, the almost two-meter-long tails can get entangled in the undergrowth. There must be something really valuable about those feathers.

Scientists have made some reasonable suggestions about why male birds are so colorful and why they sing so much, and why females choose the most colorful and loquacious males. One possibility is that the extravagant plumage of male birds demonstrates that they are healthy—well fed and, perhaps most important, free of parasites. A sick male bird cannot produce brilliant plumage. Another possibility is that the males are just showing off—showing that they are so endowed with resources and virility that they can waste it on making showy feathers and on composing symphonies. The male is also showing off by boldly risking a brush with death at the claws of a predator.

This is sexual selection: the females choose the males with the most expensive ornamentation, not because of the ornamentation itself but because of what it indicates about the health and wealth of the male. Bird plumage thus functions as what scientists call a "fitness indicator." In order for sexual selection to occur, just as for natural selection, there must be variation in the population. Indeed, some males are more colorful, and some females more choosy. Sexual selection rewards the colorful males and the choosy females. In some cases, the male birds gather at a certain location and strut around for the females to watch. Such a gathering is called a lek.

In many bird species, since the females are choosing rather than competing, they are less colorfully ornamented. As I mentioned previously, male painted buntings are green, red, and blue. The females are just a dirty green color. Female cardinals are reddish-brown. Peahens and female birds-of-paradise do not have tails as long as those of males.

Sexual selection is not the only reason for bright coloration, of course. Some of the most brightly colored amphibians are the poison-dart frogs, whose skins are highly toxic, a fact that is advertised by their brilliant color. But sexual selection has probably been the major process by which wastefully beautiful animal characteristics have evolved. Efficient natural selection would not do this. Still, what is the point of gaining lots of resources, as natural selection enables you to do, if you cannot win the game of sexual selection and leave lots of offspring?

Birds are by no means the only animals in which sexual selection produces showy males; fishes and insects provide similar examples. In most cases, males compete for access to females and are more colorful, but this is not always the case. The males and females may advertise their health to one another and choose one another by means of sexual selection. In many butterflies, for instance, males and females are more or less equally ornamented.

In some species it is the males who nurture the offspring. In such cases, the female reproductive contribution is much less: they just leave the eggs and move on. In the red-necked phalarope (a wading bird), it is the females who are more brightly colored, and they compete for the attention of the less colorful males. In pipefish and seahorses, the males take the fertilized eggs into a special pouch and brood them until they hatch. In those species as well, females compete and males choose.

A SEXUAL HISTORY OF PLANT LIFE

Plants also have sexual selection, but plants are rooted to the spot and cannot seek out mates. At first it seems that they must leave their selection of mates up to the vagaries of the wind or to the whims of pollinators: the pollen gets carried to whichever flower the pollinator decides to visit, and the plants have no choice in the matter. But is this true? Within a species, the plants with bigger flowers, or those that offer more of a nectar reward, might receive more visits from pollinators than the plants with smaller flowers. Plants with bigger flowers, then, may crossbreed with one another more often than with smaller-flowered plants of the same species. Looking at it this way, you might conclude that every time you see a pretty flower, you are seeing the product of sexual selection.[10]

Most flowers produce both pollen and seeds. But about 10 percent of plant species have separate male and female individuals (figure 5-2). One famous example is the date palm. Thousands of years ago, the Mesopotamians noticed that some date palms produced dates, and others just produced what we now know to be pollen. And they discovered that when they planted their orchards, they had to have at least a few of the pollen-producing trees. As with animals, we might expect that competition between male flowers to attract pollinators might be more intense than the competition between female flowers. In fact, in several species of plants, male flowers are larger and more colorful than female flowers.[11]

Figure 5-2. Cottonwood trees (*Populus deltoides*) are male and female. Male trees (left) produce male flowers that release pollen into the air, while female trees (right) produce female flowers on which the pollen lands, in early spring. Photographs by the author.

But this is not the only way that plants participate in sexual selection. As I mentioned previously, in connection with migratory bird species, the males arrive on the scene earlier than the females and carve out their territories. My students and I have data that suggest that a similar phenomenon may occur in plants as well. In a survey of over a hundred cottonwood trees in southern Oklahoma and north Texas, we found that the flowers of the male trees began to shed pollen into the air about a week before the flowers of the female trees opened. (Cottonwoods allow the wind to carry their pollen and, later, their seeds.) I have also noticed a similar pattern in bois-d'arc trees and holly bushes, but I did not obtain a large enough sample size to draw a firm conclusion.

But, you may ask, what possible advantage could there be for male trees to release their pollen into the air a week before there are any female flowers to fertilize? The week's difference, however, is just an average. Once a male tree begins to release pollen, it continues doing so for at least a week and a half. The male trees that open their flowers early may end up wasting about a week's worth of pollen, but the males that open their flowers later, when the females first become receptive, will find that the environment is already saturated with the pollen produced by the early male trees. It appears that male cottonwood trees are competing with one another to pollinate female cottonwood trees, though not consciously. There is some risk associated with buds opening too early. Early budburst might allow the buds to be damaged by frost. Natural selection will operate against male cottonwood trees that open their buds too soon, but sexual selection makes it worth the risk to open them as soon as possible.

SEX MAKES THE WORLD INTERESTING

Most of the characteristics I have described thus far in this chapter are not entirely the product of sexual selection; they almost always have some utilitarian purpose as well. For example, feathers serve at least two utilitarian purposes. Down feathers keep birds warm when they are young and during cold weather, and flight feathers allow birds to fly. The role of sexual selection was to add length and color to feathers that had already been brought into existence by natural selection.

Sexual selection has, therefore, filled the world with beautiful things, such as color—whether of birds or of flowers—and song. It has also filled the world with dramatic violence. What a different world it would be if there were no sexual selection, and plants and animals devoted all their resources to growing large and efficiently producing offspring without sex! It would be crammed full of plants and animals, perhaps all of them sick and squirming with parasites. But it would be very quiet and dull. And there is no guarantee that a world without sex would be less violent. Animals might just find something else to fight over.

HUMANS, THE SEXIEST SPECIES

Perhaps the species that has undergone the most extensive sexual selection is our own. How have humans been sexually selected? Let me count the ways!

As with nearly all other animal species, men can sire more children than any one woman can bear. In the ancient world, conquerors were famous for leaving hundreds of offspring (something that modern conquerors, such as Adolf Hitler, have not generally done). Nobody is sure how many children Genghis Khan may have had. In western Asia, almost 1 percent of men today carry a Y chromosome that has been traced to Genghis Khan or his immediate family.[12] The record number of children for a woman to bear is sixty-nine, born to a Russian peasant woman in the eighteenth century—a large number, but fewer than the hundreds of children sired by some males. Therefore, as in other animal species, men compete for access to women more than women do for men. This competition can take many forms, some destructive, some creative, as we will see in the following sections.

However, as we will discover, women also compete for access to men. Women are often attracted by, and compete for access to, men with strong muscles and broad shoulders. However, physical attractiveness is only one factor. Henry Kissinger, secretary of state under President Richard Nixon, said that power is an aphrodisiac. He knew a thing or two about power and about women, and his personal appearance was apparently irrelevant to the outcome.

As Darwin pointed out, sexual selection usually requires not just male combativeness but female choice. In the following sections, we will see that sexual selection acts upon both men and women.

The Naked Ape

Let us start with the most obvious feature. Humans are (in the words of zoologist Desmond Morris) naked apes.[13] Human skin has about the same number of hair follicles as the skin of other apes, but these hairs do not grow as much, or at all, over large areas of skin. The leading explanation for human hairlessness is that it allowed our ancestors to more efficiently radiate heat from our bodies, allowing us to keep our bodies from overheating on the African savanna millions of years ago. This was especially important for our large brains, which produce a lot of heat. While there are many other, and very hairy, species of mammals on the savanna, scientists claim that human hairlessness allows the human body to remain cool during long-distance running, which humans can do better than other savanna species.[14] The other savanna mammals also had smaller brains. This is the natural selection explanation for our relative hairlessness.

But sexual selection has also been proposed as a reason for human hairlessness. Some have suggested that the reason humans are apparently hairless is that men and women can check each other out for signs of illness or parasites, indicated by the health of the skin. It has also been suggested that bare skin is more erogenous than furry skin, an idea that cannot be tested. As with many other animal traits, sexual selection may have taken advantage and modified a characteristic that had already come into existence by natural selection.

Another mystery is why humans have thick hair only on their heads, in their armpits, and in their genital areas. Some scientists have said that these hairs help to waft sexual odors into the air from glands in the armpits and groin.[15] While the existence of human sexual odors has been experimentally demonstrated, and is beyond doubt, it is questionable whether the axillary and pubic hairs play any role in their effectiveness. The answer to the mystery may simply be that the head, armpits, and groin just happened to be the areas in which hairs first develop in embryos. Perhaps humans retained an embryonic pattern of hair development.

As a matter of fact, humans retain many embryonic and juvenile characteristics into adulthood. The retention of juvenile characteristics into reproductive adulthood is known as *neoteny* and is one of the most important characteristics that distinguish humans from our closest ape cousins. Humans are reproductively mature juvenile apes. We look like it, too. Ever see a picture of a young chimpanzee? It doesn't look all that different from a human being, whether a human child or a human adult. In contrast, an adult chimpanzee does not look very human at all. Juvenile chimps are less hairy than adults. Among the neotenous characteristics of humans is the size of the brain. In most animals, juveniles have larger heads relative to the sizes of their bodies than do adults of the same species, but the brain stops growing long before adulthood. During human childhood, the brain keeps growing and developing throughout the juvenile period. The large human brain is a neotenous characteristic.[16]

And, according to some scientists, women are more neotenous than men. Women have more neotenous characteristics—big eyes, soft skin—and maybe this is why men call them "babes." In most societies, the women who are considered most beautiful are those that look the youngest, while the men who are considered most desirable do not necessarily look young. Older men may be more experienced and have more resources, which is what many women want. But younger women have a longer reproductive life ahead of them, and this may be an underlying evolutionary reason why men often prefer younger women, while women often prefer older men. The underlying evolutionary reason may, of course, be very different from the psychological reasons that determine the preferences of any particular individual man or woman.

Curves

Women have a greater amount of body fat than men, even if they are skinny; much of this fat is subcutaneous, which gives them smooth skin. There is a practical value to these fat deposits: they provide extra metabolic energy that allows women to bear children and to lactate. These fat reserves also serve as fitness indicators to men, showing that the woman will be able to bear children and to lactate. Fat accumulation in women is just as much a fitness indicator as are colorful feathers in male birds. A plump woman is obviously one who has access to plenty of food. Back in the Stone Age, fat women were hot. There is a fair amount of Paleolithic pornography to this effect. It is only in recent centuries that slim women came to be admired in society more than fat women—and even then, fat accumulation in certain highly visible areas continues to be admired. Sexual selection favors women with neotenous characteristics and fat deposits, and men who desire them.

Violence

Competition among men for access to women has promoted the evolution of violent behavioral tendencies, just as in males of many other animal species. In prehistory and throughout recorded history, men fought not just over women but over any resource or status that would allow them to attract and support women. This means that they would fight over almost anything. It is the men, and seldom the women, who fight over territory, resources, and ideas. Fighting and wars have been an almost exclusively male activity. Armies of men from one city-state would conquer another city-state, enslave or kill the men, and enslave and rape the women.[17] Violent competition among men is arguably the worst part of human history.

Even though men fought one another for resources and opportunities that would increase their access to women, they did not necessarily esteem the women. In some tribal cultures, women held positions of respect. For example, the Cherokee tribe allowed women to make important decisions in village council meetings. When my Cherokee ancestor Attakullakulla visited England in 1730, he was amazed that the British did not consult their women in making important decisions.[18] But in Western cultures, women rarely held positions of esteem even into the nineteenth century, except in rare cases when a queen grasped power. Women were a resource, and men treated them as such, effectively buying and selling them through dowries and arranged marriages. The abuse of women by men was not considered as serious as even a property crime committed by one man against another. The cumulative weight of suffering of women through the ages is as incalculable as it was universal. Even into the twentieth century, in some American states (such as my state of Oklahoma) husbands had the right to make decisions about the property of their wives.[19]

Domination of women by men often had a violent and demeaning aspect to it. Female genital mutilation of girls, a crime whose horror is difficult to surpass, remains common in some parts of the world. This process makes sure that girls grow up into women whose sexual activities can be controlled. It was men who invented the metal chastity belt with lock and key. For millennia, men have used women as a resource for sexual pleasure while pretending to be holy, such as by both patronizing and condemning prostitutes. Though male gorillas and elephant seals are not known to be particularly polite to their harems, they seem angelic in comparison to the more notorious of men.

Fortunately, sexually selected male violence is only a tendency, not an inevitable practice. The tendency undoubtedly has a biological basis; therefore, it is likely that boys will always grow up to be more combative than girls, even in the most egalitarian of societies. (Remember the story of the Greek warrior Achilles. As a boy, he was hidden in a house of girls, but Odysseus found him by placing a sword among the jewelry that he offered to the girls, and Achilles immediately grabbed it.) But aggressive male tendencies can be reduced and channeled into more constructive activities, and oppressive male practices can be unlearned when new generations of men grow up in a society that treats women with respect.

Hunting and Sports

Two examples of sexually selected human behaviors are sports and hunting. Since I am a hiker and a presser of wildflowers, I have little interest in either of these pastimes, but I acknowledge their importance to the human species throughout its evolutionary history and in all societies. The practical (nonsexual) explanation for them is as follows: sports allow little humans to practice the arts of hunting and war, without actually getting hurt, in most cases. Sports are also a valuable social practice that may encourage bonds of friendship within tribes and perhaps release tensions between tribes that might otherwise turn into open conflict. Here in Oklahoma, the rivalry between the maroon and the orange football teams provides an outlet to hostilities that might otherwise find dangerous expression. And hunting is important because it brings home meat, a valuable source of protein, to feed the family. Even in the United States, some poor rural families get much of their meat from hunting.

But some evolutionary scientists, such as Geoffrey Miller, have a different explanation.[20] Sports and hunting do, of course, have the above advantages. But big-game hunting, as it turns out, was not a particularly good way of providing protein to one's family in the Stone Age. In tribal cultures, most of the calories come from the fruits, nuts, and roots gathered by the women. And the most efficient, and certainly the safest, way to bring home meat is by catching small animals. Hunting big game—especially if you do it the way Neanderthals did, by jabbing them with spears—is dangerous.[21] But, asks Miller, that's the point, is it not? In prehistoric times, men engaged in sports and hunting to show off their power and skill to the women. While the women may not have actually seen the men hunt, they certainly saw the animal carcasses they brought home and heard the stories of hunting prowess. In this way, sports and hunting are not all that different from war, in which men may needlessly engage in order to demonstrate their strength and bravery. (Apparently the first "Stanley" was a warrior or hunter. The Slavonic origin of the name *Stanislaus* means "glory of the camp"—a tradition I have not carried on.) The most glorious male would get the top pick of the chicks. That is, the women would use hunting skill as one of several indicators of the possible fitness of the men they would choose. Even when hunting in nomadic societies was replaced by herding in settled cultures, males continued the tradition of making it into an adventure, as any of my rodeo-loving Oklahoma neighbors can tell you. It is rare for men to compete with one another by bringing home vegetables, although I have heard that in New Guinea, the men compete to see which of them can bring home the biggest yams. And in New Guinea, the yams get pretty big.

Sexual selection has therefore resulted in strong men who like sports and hunting (and war), and women who admire such men. As is always the case with traits that have been influenced by evolution, there is variation. Not all men like to do these things, nor do all women admire them for it. As I will explain later, there are other, very different ways that men have competed for the admiration of women. Furthermore, sexual selection can operate both ways with respect to hunting and sports. Many women are very good at sports or hunting (à la Sarah Palin), and men may admire them greatly for it.

Sex and the Evolution of the Human Mind

The most important product of human sexual selection and neoteny is the human brain itself. Humans are grossly intelligent. The size of our brains, relative to our bodies, exceeds that of any other species by a wide margin. Intelligence is expensive: the brain uses a titanic share of the sugar and oxygen carried by the blood. Why do we have it?

To answer this question, we must consider what intelligence is like. If intelligence were mainly to solve problems—to figure out how to get food or to build shelter or to outwit enemies—then the human brain would be logical. Logic and reason would come naturally to us. We are indeed capable of logic, but usually we have to work at it. Logic is not what the brain does most easily. This becomes even more apparent when we consider the most outstanding product of human intelligence: language. If the main function of language over evolutionary time were to convey information, the structures of languages would be logical and efficient. This is far from being the case. In fact, the most primitive languages are the most complex, with many different noun declensions and verb tenses, and many characteristics that you might not understand even if I understood them myself enough to explain them to you (such as numeral classifiers). Ralph Waldo Emerson thought that primitive languages such as those of Native Americans were simple, like the words of children, but it is clear that he did not bother to talk to anyone who spoke or had studied those languages.[22] The complexity of Navaho makes English sound like children's talk. For perhaps most of human prehistory, languages were as grossly complex as the human brain. Intelligence, and language, did not evolve to solve problems or to explain things clearly.

The most important answer to the riddle of human intelligence is this: the human brain, with its language loudspeaker, is an entertainment system. Prehistoric people, especially when they were telling stories around the campfire, did not want to speak clearly and simply. They wanted to speak beautifully. And those who could tell stories, or sing, or dance, or paint caves (as at Lascaux in what is now France), or produce sculptures (as at Dolní Věstonice in what is now the Czech Republic) most beautifully were considered to be the most attractive and the most godlike. This would be true for both men and women. The human brain got big enough that we tried, and failed, to understand the world, and religion (as will be explained in chapter 7) was our way of dealing with it. The human brain wanted to make a hostile and confusing world into a beautiful cosmos that could be understood.

Therefore, men who were most attractive to women were not just the strongest, they were also the most intelligently artistic. And the women who were most attractive to men were not just the most curvaceous but also the most intelligently artistic ones. It was not just the great hunter and the fat babe who were most sexually attractive in prehistoric times, it was also the wise storyteller and the keeper of the tribal lore, whether male or female. There were probably Stone Age geeks as well as Stone Age hunks who got lots of reproductive opportunities. At least that is what I like to think. The human brain was not just an entertainment system; it was a sexually selected entertainment system. This is why humans have never been satisfied with being just as intelligent as was necessary to survive.[23]

Sexual selection is not the only reason for language, of course. Another factor that contributed to the evolution of intelligence and language was the ability to keep track of allegiances and loyalties within the tribe.[24] Primitive languages are complex also because they served as identifiers of culture. Only native speakers of a language could master its complexities; outsiders could not fake it. This is the idea behind the "shibboleth," a word that a nonnative speaker cannot pronounce. It comes from the Old Testament story of a word that Gileadites could pronounce but Ephraimites could not.[25]

One of the most important components of human intelligence is humor. Humans, like most mammals, can be happy, but humor is the intellectual dimension of happiness that is peculiarly powerful in humans. Humor is the enjoyment of incongruity. There is no particular reason why an intelligent being should enjoy incongruity; computers certainly do not. Like all other aspects of intelligence, humor serves many purposes. It creates social bonding and can allow aggression to be handled less violently. It is a human way of dealing with environmental adversity; humor can help keep people alive through catastrophes. But it may be primarily the product of sexual selection. Humor is a universal component of creativity. A tribe admires the man or woman who can get them to laugh, and this person can also get his or her prospective partner to laugh and love. Humor can lead to intimacy because the person creating it becomes vulnerable, making fun of himself or herself a little. Statements that are purely derisive and demeaning, unmixed with the speaker's humility, are not funny to most people.

Sexual Selection and Cultural Extensions

Cultural and technological extensions of an animal's body also function in sexual selection. Male bowerbirds collect blue objects, including blue garbage, to create inviting little tunnels of love to lure their mates, rather than growing immense blue tails. Humans use culture and technology as extensions of bodily characteristics more than any other kind of animal. Humans often use displays of wealth as fitness indicators. Clothes, houses, and vehicles function as sexually selected technological extensions of the human body.

Cultural and technological extensions are not always reliable fitness indicators, however. Wealth may be a good fitness indicator if the man or woman got it by their own actions. But inherited wealth is a famously poor indicator of mate quality. The story of the girl who elopes with a poor but able young man rather than marrying the rich heir has been retold thousands of times in many cultures. Many young males like to drive around in loud, powerful vehicles such as pickup trucks with oversized tires or the aptly named "muscle cars." When women or men find themselves swayed by cultural or technological extensions of the body of a member of the opposite sex, the effect is often subconscious. While a Ferrari may be a good indicator of wealth, a big pickup truck (complete with truck nuts) bought on credit from an American auto maker facing bankruptcy may not be. No young woman would *consciously* tell herself when she sees the guy in the pickup truck, "Wow! He must be quite a guy! He can push down a gas pedal!" Sexual selection has acted on cultural and technological fitness indicators, as imperfect as they are.

Cultural and technological enhancement of sex appeal was undoubtedly occurring in prehistoric times as well, though the evidence is less clear. Much of what we know about prehistory comes from the study of stone tools. Stone tools were not made just to be practical; they were also made to demonstrate the knapping skill of the toolmakers. That is, prehistoric women might have found men who could make good stone tools to be attractive (assuming that it was men who made the tools). This may be the only reason that, a million years ago, during the time of Acheulean technology, men made some hand axes that were much too large to use (figure 5-3).

Figure 5-3. Ian Tattersall (American Museum of Natural History) holds a replica of one of the gigantic Acheulean blades that *Homo ergaster* made in Africa about a million and a half years ago. Too large and heavy to actually use, these tools may have served an artistic function.

(*Acheulean* comes from the Greek for "before truck nuts.")[26] Then as now, however, cultural and technological extensions were imperfect indicators. Tribes often produced only a limited range of tools, and they kept making them the same way for many thousands of years. One location in East Africa had a tool technology that persisted unchanged for a million years.[27] This kind of cultural stagnation is not the way for men to impress women. As evolutionary scientist Matt Ridley indicated, a prehistoric woman was not likely to say, "Oh, honey, just what I always wanted—another hand axe."[28]

Sexual Selection and the Origin of Races

Darwin proposed that sexual selection was the main reason for the existence of human races.[29] It is true that natural selection explains a few things about the differences among races. For example, dark skin is advantageous in the tropics (where tribal peoples traditionally did not wear clothes) because the melanin absorbs ultraviolet radiation and helps to prevent mutations in skin cells, which lead to cancer. In high latitudes, where people wore clothes except perhaps in summer, light skin is more advantageous. Melanin would block the sunlight that is essential for the manufacture of vitamin D in the temporarily exposed skin. Aside from this, there do not appear to be any good adaptive or survival reasons for racial differences. Instead, races evolved as isolated populations, and each population had its own arbitrary idea of beauty. This very arbitrariness is a hallmark of sexual selection: it has produced human racial variety just as it has produced great variety in the colors of bird feathers.

Racial distinctions evolved by sexual selection in prehistory. These distinctions were not great enough, however, to impair our ability to recognize all humans as a single species. When humans began to travel extensively, and races came in contact, they immediately began to mate with one another.

In nineteenth-century Europe and America, scholars debated whether the different races were or were not a single species. Abolitionists claimed all humans were members of a single species. Religious abolitionists cited Adam and Eve as their reason for this belief; Darwin's proposal of the evolutionary origin of all races from a common ancestor grew from his family's strong abolitionist beliefs. Predictably, slave owners and their scholarly defenders claimed that dark races were created separately and were inferior.[30] But even as this debate went on, black and white people interbred with one another, and slave owners and racists had to invent arbitrary rules to deal with this inconvenient fact. The "one-drop rule" said that "one drop of black blood" made you black. The absurdity of this viewpoint was shown by Mark Twain with his characteristic wisdom in his novel *Pudd'nhead Wilson*, in which a white baby and a baby that was one-eighth black were switched in their cribs.

Many of us in the United States are the descendants of mixed marriages, especially between white and Native American, as I am, or between white and black, such as President Barack Obama. We all know that human races are a single species, and those of us who have loved a person of a different race know it very personally. Some scholars predict that races may homogenize during the next few centuries, as is already happening in Hawaii and Brazil.[31] Already, images of sexy female whites, blacks, and Asians all look like Spice Girl cartoons.

Monogamy, More or Less

Gorillas have harems, as do some human cultures. Chimps are promiscuous, as are some groups of humans and many individuals. But most humans have a cultural norm of monogamy—or, at least, of serial monogamy, having one partner at a time, often long enough to raise a child. Human offspring are so expensive to raise, especially if they go to college, that it is best done by two parents. It's hard enough being a single mom today; imagine what it was like in prehistoric times! As explained previously, a man's evolutionary fitness is not enhanced by siring a child who dies. Some men could force or entice other people to take care of the children. But for every Genghis Khan or Don Juan, there are at least a few, and perhaps a few dozen, men who cohabit with one female for a long period of time and enjoy it—not merely out of moral conviction but because evolution has reinforced the feelings of love that promote the successful production of children. Perhaps, you see, the best measure of evolutionary fitness is not the number of children but the number of grandchildren; you are successful in evolutionary terms when you become an ancestor.

Not surprisingly, then, sexual selection has produced in humans a number of adaptations that encourage a man and a woman to stay together. One of these, according to some evolutionary scientists, is concealed ovulation. In many mammals, including chimps, a female goes into a period of estrus when she ovulates about once a month. Her sexual organs swell, and she releases a scent irresistible to males, who follow her around and all try to mate with her at once. Dogs do this also. It's a wild and crazy time when the bitch is in heat. But human females, as well as those of some other species of primates, ovulate without producing any visible evidence of it. Some women claim to know when they are ovulating, while others scoff at the idea. Women become more receptive to indicators of male attractiveness, whether appearance or hormones, during ovulation, but this process is subconscious.[32] The only thing women know for certain is when they are about to menstruate. Before people wore clothes, everybody knew which woman was menstruating—she was the one sitting on a little pile of straw. But for three weeks each month, ovulation could occur at any time, and neither the male nor the female would be likely to know, even after it had occurred.

As a result of concealed ovulation, according to this view, if a man wanted to make sure that a woman's child was truly his, there was only one way he could do this: by staying with her throughout her nonmenstrual time, keeping other men away. For a man, a quick copulation and a quick departure is not a very good way to produce a child with a woman whom he has chosen. If he copulates promiscuously often enough he will probably produce a child, but he would not know which child is his so that he could provide it with resources and protection. According to this theory, sometimes called the "daddy-at-home" theory, concealed ovulation may be an evolutionary mechanism for keeping men around at least for a while.[33] Perhaps, after the woman's first missed period, the man can run off. But since miscarriages were even more common in prehistoric times than they are today (about one-third of all pregnancies end in spontaneous abortions), he might want to stick around longer. And he might stick around for many years, to provide for his children. The result is partially or completely monogamous pair bonds.[34]

Some evolutionary scientists such as Sarah Blaffer Hrdy suggest a more ominous reason for concealed ovulation. In some species of mammals, infanticide is common. Male lions frequently take over female lions, and if the females already have cubs, they kill them. The female lion then becomes ready to bear his cubs. Infanticide is much rarer in humans, but it is still disturbingly frequent. It may have been even more common in prehistoric times. If several males mate with a female during her three nonmenstrual weeks, and the female becomes pregnant, none of the men can be sure which of them might be the father. As a result of this uncertainty, all the men refrain from killing the child. According to this theory, sometimes called the "many fathers" theory, concealed ovulation might have evolved as a way of preventing infanticide.[35] This explanation makes sense in a relatively promiscuous mating system—but also in any situation where the man cannot be entirely certain of the woman's fidelity to him.

LOVE

Yet another evolutionary adaptation that helps keep men and women together is love. And now you get to read a few paragraphs in which a science geek tries to explain the wonders of love! Lucky you!

What is love? Throughout the ages, this question has been answered in poetry rather than fact. Even the otherwise confidently theological Apostle Paul devoted part of his first letter to the church at Corinth (which we recognize as chapter 13) to beating around the bush and never quite saying what love is. Love does this, or is like that, but nobody can say what it is.

Kinds of Love

The answer, of course, is that there are many kinds of love. The Greeks recognized at least four kinds of love. The ancient Greeks did not necessarily know anymore about love than you do, but their findings at least give us a conceptual framework with which to begin.

One kind of love is *storge*, which is a mother's love for her children—and, to a usually but not always lesser extent, a father's love. This kind of love is well known from the animal kingdom. A mother bird may pretend to have a broken wing and lure a predator (or a birdwatcher she thinks is a predator) away from her young. The sacrifice that mother animals are willing to make for their offspring is one of the most powerful forces of nature.

The *storge* kind of love often shows up in situations other than the love of parents for their children. It is particularly common in religious art. Angels are often depicted with childlike features. One particular artist in the American Midwest produces paintings and sculptures in which the child-angels have large, teardrop-shaped eyes, with a tiny nose and mouth represented by little parentheses near the bottom of the face. Some people spend hundreds or thousands of dollars collecting this artist's paintings and figurines. This artwork draws its appeal from a straight shot of two-hundred-proof *storge*.

Two other kinds of love are *philia*, which is the love between friends, whether of the same or of the opposite gender; and *agape*, which is gentle, unselfish, spiritual love. None of these three kinds of love is overtly sexual. The Greeks, of course, recognized a fourth kind of love, *eros*, or erotic love.

Eros

Even erotic love is not a single thing. It is at least three different feelings, each one of which has its own set of brain chemicals and areas of brain activity. Anyone who has been in love has experienced all three of these feelings and will recognize this description of them.[36]

One component of erotic love is, of course, lust, which I will briefly describe from a masculine viewpoint. When a heterosexual man sees a buxom and curvaceous woman, especially when she is partly or wholly nude, lust overloads all his circuits. Depending on the context, the woman does not have to be particularly buxom, or particularly curvaceous, or even particularly nude. In repressive societies, where women are kept bundled up, let her show so much as an ankle, and a man gets hot. (I own a late nineteenth-century schoolbook that the adolescent male student filled with sketches of women in skirts with large whalebone bustles.) It is, in fact, an ankle that inflames the lust of the main character in the memorable final scene of the Sinclair Lewis novel *Elmer Gantry*. In societies where nudity is common, it is at least a little less stimulating. Often, nudity is not enough: women sometimes adorn themselves with special gear to make themselves more alluring to certain kinds of men. One of the most memorable scenes in the 1985 adventure movie *Emerald Forest* is when the Amazonian tribal women who have been kidnapped and dressed in sexually stimulating clothes are rescued. They run out into the forest and tear off the bordello clothes in disgust, returning to the relative decency of their erstwhile nudity.

Lust is driven largely by testosterone (in men as well as women). A man can lust after many women, and women after many men, at the same time. What evolutionary good does lust do? It stimulates a man or woman to pay close attention to the field of possible sexual partners and to work very hard to choose one.

The second component of erotic love is the actual passion itself: being in love. Lust can have many objects at the same time, but a person is in love with only one other person at a time. When you are in love, you are euphoric when things are going well, and when they are not, you have a sinking feeling of desperation. You are utterly obsessed with and powerfully crave that one individual. But even being with the person is not enough. You want to smother the other person in your intense emotion. The object of your desire seems to embody every pleasing sight, sound, taste, scent, and touch, and yet even when you are with the other person, you feel subjected to maddening blindness, deafness, and the inability to taste, smell, or feel. This emotion is one that is entirely without satisfaction. Even if the orchestral music swells and the significant other says, "Yes! I love you too!" you cannot be completely happy because you are worrying about whether this person will love you in the future—that is, approximately two minutes from now.

Being in love is what happens when the brain has high levels of the neurotransmitters dopamine and norepinephrine and low levels of serotonin. Yes, a few simple chemicals can make all this happen in your brain. Dopamine gives you a sense of exhilaration, accompanied by sleeplessness, trembling, and a loss of appetite. It is associated with drives, rather than with milder emotions. Norepinephrine focuses your attention and memory on every detail of your beloved. Low levels of serotonin are associated with obsession. I am amused that dopamine makes you feel like a dope or as if you are high on dope. I have not had this feeling for a long time, and I'm sure many of you will join me in saying, *Good riddance!* Falling in love really is a type of falling—down into a bottomless pit. In terms of brain chemistry, it is an addiction.[37] It was immortally expressed by the writer of the Song of Solomon: "I am sick with love." Being in love is as effective as a virus in sapping a person of strength and sending him or her to bed (alone). Once, a long time ago, I was in unrequited love and had the flu at the same time, and I leave you to guess which one had more of an impact on my health.

But the evolutionary function of this passion is clear. When you are in love, you will stop at nothing to protect, defend, and nourish the object of your desire. You will go berserk in battle or run a hundred miles until you fall, exhausted and bloody and muddy, at his or her feet. What concealed ovulation and even lust cannot do, this feeling can. And very few human beings have escaped its absolute, though thankfully brief, tyranny. It can even be considered a temporary insanity. This is the primary reason that unrequited love is so painful, and an unwanted lover is such an annoyance and possibly a danger.

But when the effects of dopamine have worn off, there is the third aspect of erotic love, which is purely pleasurable and outlasts the others. This is the deeply intense feeling of companionship and attachment. It feels like heaven, but it is just oxytocin and vasopressin. Testosterone and dopamine can start a relationship, but it is oxytocin that keeps it going, often for many years. Although old people are capable of lust and of falling in love, it is oxytocin-induced bonding to which they refer when they say, "We've been in love for fifty years." No one could be high on dopamine for fifty years. But oxytocin can do it.

All these hormones and neurotransmitters interact with one another, and none of them has a single, simple effect. This is why erotic love is so complicated. Erotic love is also a specific product of a lifetime of experiences within a culture and can thus hardly be the same for any two people. But, in general, testosterone makes you scramble up the loose scree of a mountain, dopamine makes you yodel at the top, and oxytocin makes you sit and smile from the high plateau.

SEXUAL SELECTION AND THE BEAUTY OF THE WORLD

Such has been the effect of sexual selection on the world of species. It has made the web of life on Earth ravishingly beautiful. And such has been the effect of sexual selection on the human species. Natural selection could not have done this; clearly, the sensible thing to do would be to objectively choose a suitable copulatory partner and produce offspring, which you would then provision well, like the stereotyped Victorian or Edwardian father. But without sexual selection, the human species would not have any of the things that make our cultures beautiful and thrilling. We would have had intelligence, but of some other kind, perhaps like the Vulcan people on *Star Trek*. There would be no crimes of passion but probably plenty of other kinds of crime. Language would be deathly literalistic. There would be laughter. But what kind of laughter would it be?

Chapter 6

ALTRUISM

It feels good to be good.

Hardly any other proof is needed that conscience and morality, found in the minds and behavior of all people, are the products of evolution. We are the evolutionary descendants of social apes who were good to one another, at least some of the time, and enjoyed it. We like the taste of fruits because our vertebrate ancestors got their vitamins by eating them. The animals that sought out fruits most vigorously were those that liked fruits the most—and natural selection favored them. The result is animals that think fruits are delicious. We like the taste of meat because our vertebrate ancestors got their protein by eating it. The animals that sought out meat most vigorously were those that liked it the most—and natural selection favored them. The result is animals that think meat is delicious. We like sex because we are the descendants of animals that reproduced efficiently and enjoyed it. The result is too obvious to state. Natural selection favors the enjoyment of beneficial things. We can be pretty sure that if being good did not provide evolutionary advantages, we would not enjoy it. But, with the exception of psychopaths, we all have consciences and enjoy being good, at least most of the time.[1]

ALTRUISM AND EVOLUTION

Altruism occurs when animals act in a way that benefits another animal of the same species. The altruist suffers some kind of loss—of resources, of opportunities, perhaps even of life—by doing this. Humans are the most altruistic species that has ever existed on the Earth. The roots of altruism go far into the evolutionary past, but the full emergence of altruism came with the evolution of humans. Since human activities are now transforming the entire Earth, altruism is also one of the most important processes in the life of the planet.

"Survival of the Fittest"

Altruism might seem puzzling at first. Doesn't natural selection operate by means of "the law of the jungle" and "survival of the fittest"? Don't nice guys finish last? Doesn't sexual selection operate because animals will do whatever it takes to get a mate? Don't nice guys finish last in the bedroom as well as on the battlefield and in the market?

Many religious people believe that ruthless competition and exploitation is the way of the world, and nothing except their brand of religion will stop it. They believe that all humans would be fighting and raping one another all the time if it were not for religious morality that is consciously embraced, conferred by a Holy Spirit available only at their church.

This is, however, a deceptive lie. The evidence is that altruism and religion (defined as adherence to doctrinal beliefs) are not consistently associated with one another. This is true on both the national and individual levels. First, the more religious countries, such as the United States and Islamic nations, do not appear to have greater levels of altruism (as reflected in lower crime rates or higher rates of philanthropic donations, for example) than do less religious countries, such as Japan or the countries of western Europe. In particular, Christian America is no beacon of social goodness.[2] Second, the more religious individuals do not appear to have greater levels of altruism than do less religious people. Even people who are not particularly religious behave in a moral, altruistic fashion and will rush to aid a fellow human in distress. Often the unreligious people are more moral. A famous psychological study provided an experimental demonstration that conventionally religious people were no more likely than other people to help a stranger.[3] Participants in this study went to a psychology lab for documentation of their religious beliefs. On their way to the lab, the participants encountered people who, as part of the experiment, pretended to need help. That is, the opportunity for altruism was experimentally imposed on the participants. The participants who were later self-identified as religious were *less* likely to have stopped to help. Perhaps this was because the religious people felt that, by being religious, they had already discharged their altruistic duty, whereas the nonreligious people had no other outlet for altruistic feelings other than to be altruistic.

Everyone, except psychopaths, is altruistic to some extent. In some people, those we regard as evil, this extent is very small. This is also a hallmark of evolutionary origin: within the population (in this case, of humans), there is variation for the trait—all degrees of altruism are found.

Some people pretend to be more altruistic than they are. This, too, demonstrates an evolutionary origin of altruism. Altruism must provide an evolutionary advantage in our species, or else nobody would bother to fake it. The strategy of faking altruism, however, does not work very well. It is actually quite difficult to pretend to be good without letting the secret core of evil show through. The easiest and most effective way to appear good is to actually be good—and to enjoy it.

Religious people are no less likely to be faking their altruism than nonreligious people. Many people use religion as a screen to hide their evil behavior. Preachers advertise their churches as the only safe haven for goodness in order to get people to join and become "giving units," the standard church term for families that make financial contributions. Sometimes, religion helps people to be altruistic, and sometimes not. After all, organized religion has been one of the principal motivations for human cruelty. In fact—as will be explored in the next chapter—religion might even be parasitic upon altruism.

But the problem remains: how could altruism have evolved by "survival of the fittest" evolution? That is what this chapter is about.

Altruism is not selfless. It is, in fact, a disguised form of self-interest. And since evolution is all about the rewarding of self-interest, it is here that we will look for the answer to where altruism came from.

"Survival of the fittest" is actually a very misleading way to describe evolution, and it was not what Charles Darwin had in mind about the way evolution works. In fact, the term "survival of the fittest" was invented by Herbert Spencer, who wrote and spoke about vague evolutionary theories and was immensely popular in the nineteenth century; he is largely, and justly, forgotten today.[4] "Survival of the fittest" calls forth images of animals fighting one another mercilessly, and the strongest prevailing. Even among plants, the stronger grow over and strangle the weaker. This viewpoint shows up sometimes in modern fiction. Kurt Vonnegut, for example, wrote in his novel *Galápagos* that Darwin "taught that those who die are meant to die, that corpses are improvements." As we saw in chapter 2, Darwin explained evolution as a process in which the most successful organisms left the most offspring. Reproductive success does not necessarily mean conquest. Organisms can serve their own best interests by quiet competition as much as by conflict, perhaps even more; and by cooperation as much as by competition, perhaps even more. We saw in chapter 4 that natural selection explains symbiosis; now we will see that it explains altruism.

Conquerors have used the bloody-conflict version of evolution as a pseudo-scientific excuse for their violence, most famously when the Nazis tried to exterminate the Jews and in Stalin's massive and cruel social experiment. But this was not even the idea that Herbert Spencer wanted to promote. His idea was that those who prevailed in the competition of life were free to be generous to the less fortunate parts of society. This was the Spencerian vision that American industrialist Andrew Carnegie embraced: he was ruthless in the manufacturing world, but then he gave away his fortune to build libraries and establish scientific research institutes, many of which still exist. Carnegie said that nobody should die rich. Therefore, even the inventor of the phrase "survival of the fittest" believed that ruthless capitalism was merely a preparation for a society based on altruism. This was similar to the viewpoint of Ralph Waldo Emerson, who in his essay "The Young American" praised American entrepreneurship but proclaimed that the communes that were beginning to take shape in the mid-nineteenth century were the next stage of human progress.[5] He did not envision this as state-sponsored or state-controlled communism but as voluntary acts of community. The experimental communes of Emerson's day failed, and we may have to wait forever for this next stage of human social progress.

We conclude, therefore, that altruism evolved, reaching a high level of intensity in the human species. And it evolved because it provided fitness or reproductive advantages to the organisms that practiced it. Our task now is to figure out what these advantages were and are. The clue that unites all the following examples is that altruism occurs in social species whose individuals are in nearly continuous contact with one another. That is, altruism has evolved in animal species for which society is the most important aspect of the environment.

"I WOULD DIE FOR EIGHT COUSINS"

In early 2009, a man in Nevada was operating a backhoe. The grinding of the engine was very disturbing to some of his neighbors—several thousand of them, in fact. But when the backhoe dislodged a boulder, this was more than the neighbors could stand, and they came rushing out of their hive. Within minutes, the man had received over a thousand bee stings. After hospitalization, he survived.

Altruistic Bees

It is well known that honeybees defend their hives from perceived threats. Africanized honeybees, which arrived in the United States from Brazil in the late twentieth century, have a much lower tolerance of perceived threats than do the European honeybees, which arrived in the United States in the seventeenth century.

When bees defend their hives, they often die. The stinger imbeds barbs into the victim's skin, and when the bee pulls away, the stinger and some of the bee's guts are left behind. Bees may give their lives in the defense of other bees. This is altruism par excellence.

Some creationists like to use this as an example of something that could not have resulted from evolution. Many years ago, I heard a creationist telling another person with intense sincerity, "Imagine the first bee that ever did this. He died. He could not pass on this trait to future generations. Therefore, it could not have evolved." Let us overlook for the moment the fact that worker bees are female. This argument is not too bad, except that, if correct, it proves that honeybees cannot exist. Within a few generations of giving their little bee lives to save their hives, the suicide genes and the bees that carry them would become extinct. Unless, that is, God miraculously re-creates the suicide genes each generation.

The worker bees (including the attack bees) are sterile females. They do not mate or lay eggs. All the offspring come from eggs laid by the queen bee, who is the only fertile female in the hive. It is the queen who passes on the suicide genes to all the offspring, most of whom will become workers. The queen, in effect, does the reproducing for them.

The queen, like most sexual animals, has two sets of genes—one set from her mother, one set from her father. She produces eggs that have only one copy of each of the genes. She mates with male drone bees, whose sperm fertilize the eggs. The queen usually lays fertilized eggs, which have two sets of genes, and these eggs become females. But she also lays some unfertilized eggs, which have only a single set of genes. These eggs hatch into the male drones. The worker bees, therefore, are all at least half-sisters, and if they happen to have the same father as well, they are sisters. The workers do not reproduce, but when the queen reproduces, she passes many of the worker genes into the next generation.

The queen appears to be in charge of the hive, but this is not the case. Most of the female eggs the queen lays develop into workers, but a small number of them, the ones that eat "royal jelly," develop into new queens. And it is the workers who decide which, and how many, of the female grubs will be fed royal jelly. The old queen may die. The new queen may take the bees in a swarm to another location, or the hive (if large enough) may split into more than one group, each with its own queen. The queens may even have a fight to the death, as the workers watch and, we may imagine, cheer them on. The queen is a trapped egg-laying machine that is occasionally used for gladiatorial entertainment.

Bees and Kin Selection

Remember that all the worker bees are either half-sisters or sisters. But the world is full of insects and spiders and fishes and birds and mammals that grow up surrounded by brothers and sisters. In few other species do the sisters give up their own reproduction to help their mother. Ants, which are closely related to bees, are another example. What is so special about bees and ants that makes them suicidally altruistic?

To answer this, we have to do a little math. Members of a species have all the same genes. But they may have two different *versions* of those genes. Two siblings may receive the same or different versions of the gene from their mother. The same is true of the genes they receive from their father. There is the remote chance that the two siblings may receive the same versions of all the genes from both parents. There is also the remote chance that the two siblings may receive different versions of all the genes from their parents. But on the average, they receive half of the same versions of the genes from each of their parents. Their genetic relatedness is therefore about one-half (50 percent). By similar reasoning, an organism has a genetic relatedness of one-quarter (25 percent) to each of its aunts and uncles, and one-eighth (12.5 percent) to each of its cousins. What this means, in terms of passing genes onto the next generation, is that your brother or your sister can pass your genes onto the next generation, but only half as well as you can yourself. Your evolutionary reproductive success is, therefore, half as good if you rely on a brother or sister to do it for you as it would be if you did it yourself. It would take eight cousins to do this for you.

And here is where altruism comes in. If you are altruistic to your siblings, you are promoting your own evolutionary fitness. If you devote yourself entirely to them, even to the extent of having no children of your own, there is no difference in the evolutionary outcome. You could even give your life to save them, without incurring an evolutionary penalty. If you give your life to save them after you have had your own children, you actually come out ahead. This is how natural selection can reward even the ultimate sacrifice of altruism. You would have to save the lives of eight cousins, however, to have an equivalent level of evolutionary success. And this is why evolutionary biologist J. B. S. Haldane, in the early twentieth century, is reported to have said, "I would die for two brothers or eight cousins."[6]

The result, therefore, is that your evolutionary fitness is not just your own but includes that of your relatives: this is called *inclusive fitness*. And natural selection does not just reward you directly but also rewards you through your kin: this is called *kin selection*.

This brings us back to the curious case of the bees. Why should sister and half-sister bees engage in such ferocious altruism? It was evolutionary biologist William D. Hamilton who figured out the elegantly simple math that answers this question.[7] Siblings of most animal species are only half related to their mother and half related to their father. But since male bees have only one set of genes, rather than two, the worker bees all receive the same versions of genes from their father. If they all have the same mother and the same father, they are related to one another not by one-half (50 percent) but by three-quarters (75 percent). Therefore, kin selection more strongly rewards altruism in sister bees than in siblings of other species. This is why bees are super-altruists. Many ants have a similar reproductive system. The reason that bees and ants are ferociously altruistic is their unusual reproductive system.

There is a price to pay, however. Entomologist Heather Mattila and coworkers have found that bee colonies in which all the workers have the same father (that is, the queen has mated with just one drone) have slower population growth than do bee colonies that have more than one father.[8] Nevertheless, the benefits of kin selection are great enough that, if a bee colony should appear that was selfish enough to not have self-sacrificial workers, it would undoubtedly lose out in competition with the efficient, cooperative, unselfish bee colonies.

Family Units and Superorganisms

There are other examples of altruism in which workers surrender their own reproduction to promote the fitness of one or a few members of the colony. Termites, for example, have sterile workers. The workers in a termite colony are all the descendants of a king male and queen female.

Most examples of kin selection altruism are less striking than bees, ants, and termites. Naked mole rats, which are rodents that dig through the soil to eat roots in Africa, allow their queens to do the reproducing. In packs of wolves, hyenas, and gorillas, dominant males or females (or both) do most of the reproduction, while the rest of the members of the pack help them raise and defend the young. Subordinate female hyenas may even go through a false pregnancy, which enables them to suckle pups that are not theirs.[9] Many animals take risks to protect other members of their group. If a ground squirrel sees a predator, it may risk its own safety by making a warning call, alerting other squirrels to seek protection. It is most likely to do this if the nearby squirrels are close genetic relatives.[10] The same is true in populations of prairie dogs (figure 6-1).

Figure 6-1. Prairie dogs are rodents that live communally in large interconnected systems of tunnels in the dry grasslands of North America. Some of the prairie dogs stand as sentinels at tunnel entrances, putting themselves in danger in order to protect the other prairie dogs, which are likely to be close relatives. Photograph by the author.

Even slime molds practice altruism. A whole colony of microscopic slime mold cells organizes itself into a single unit, producing a long stalk with a capsule at the top. Within this capsule, some of the slime mold cells produce reproductive spores that blow away in the wind. The cells that are not inside this capsule die without leaving offspring.[11] Some scientists even claim that plants may undergo kin selection. In one experiment, plants grew better when close genetic relatives were nearby than when less closely related plants were nearby.[12]

Kin selection has proven immensely successful over evolutionary time. Only a small percentage of insect species has the honeybee type of reproduction, but those species have enormously larger populations than do the species that reproduce in the typical animal way. There are many species of solitary bees and ants; most of them rare, but social bees and ants fill the air and ground throughout the world. Termites process so much wood (emitting methane in the process) that they are a major component of the global carbon cycle. Kin selection has, therefore, altered the entire planet.

The most extreme example of inclusive fitness is the cells of our bodies. Almost all our cells are genetically identical: their relatedness is 1.0 (100 percent). We never give a second thought to the fact that our cells will sacrifice themselves for the safety and success of our bodies. Among the cells of a body, altruism reigns supreme; among siblings, altruism is important; among honeybee sisters, the intensity of altruism is somewhere in between. In most animals, siblings support and defend one another but act like separate organisms. Your cells support and defend one another, acting like a single organism. Bees and ants are in between: siblings support and defend one another, acting like what biologists Edward O. Wilson and Bert Hölldobler call a *superorganism*—that is, a single organism that consists of separate bodies.[13]

Kin selection is also important in humans. Families and clans are notorious for sticking together, hence the saying that blood is thicker than water. Favoritism toward family members is nearly subconscious and occurs even when a person is trying to be objective and fair. I can in complete honesty declare my daughter to be one of the smartest, nicest, prettiest, and most socially adept people you could ever hope to hire. But this just might be because I am blind to any but the most obvious ways in which other applicants may be superior. This is why many organizations have antinepotism rules: a supervisor is not allowed to hire, evaluate, or promote a family member. Humans may be less zealous in the preservation of their kin than are bees, but we are unique in having long-distance kin selection. Our intelligence permits us to remain in contact with and work together with kin with whom we are not in direct contact.

DOING WELL BY DOING GOOD

If kin selection were the only process that produced altruism, then humans would not be particularly altruistic: they would be no more altruistic than ground squirrels but less so than bees. But kin selection is not the only way altruism can evolve. It can also evolve through *reciprocity*.

Reciprocity occurs when an animal performs some act of altruism for which repayment is likely. This repayment comes from other animals in the same species, whether or not they are close genetic relatives. Humans call it enlightened self-interest, and human societies universally recognize that there are immense benefits to be had and profits to be made when people work in partnership. Working in this way allows different people to specialize on different activities, with the result that more things get done (and done well) than would be the case if everyone tried to do everything for themselves. The economic system of every tribe and nation has resulted directly from this.

Reciprocity and Intelligence

One requirement for reciprocity to be successful is intelligence. There are always freeloaders and cheaters that take advantage of the altruism of other members of the group without being altruistic themselves. Animals that engage in reciprocity have to be smart enough to recognize and to remember who is a freeloader and who is a reciprocator. Many mammals act altruistically toward others that are behaving altruistically toward them; the most intelligent mammals can remember past acts of altruism and reciprocate later. Animals also have to be smart enough to figure out which other animals are faking the altruism. That is, they have to have what Ernest Hemingway called built-in bullshit detectors. In large groups of animals, the resulting list of friends and foes can be very complex, which is one reason why reciprocity is known only in intelligent animal species, such as meerkats, elephants, whales, dolphins, monkeys, apes, and humans. Put another way, intelligence places a limit on how extensive the network of reciprocity can be.

The development of large networks of reciprocity—within and between tribes—has been one of the greatest benefits that humans have experienced during our evolution. Without doubt, reciprocity has been one of the most important factors, along with sex, in the evolution of human intelligence, especially of language. The human brain evolved partly as a problem solver—to figure out where to find food and to invent tools for processing it—and as a sexually selected entertainment system, as explained in the previous chapter, but also as a creator of complex altruistic societies.

Intelligence has made itself indispensable in our species. We now live, and have lived for perhaps a million years or more, not so much in literal jungles as in social jungles. Reciprocity has allowed human tribes to withstand nearly every environmental condition and catastrophe, including one that, about seventy thousand years ago, nearly caused *Homo sapiens* to become extinct.[14] Reciprocity allowed humans to eventually spread throughout the entire earth. Reciprocity is the reason humans are now the species that has had a greater impact on the planet than any other species that has ever existed—much more so than bees and ants and termites that depend solely on kin selection. At the same time, any human who cannot navigate through the jungles of the mind stands little chance of being successful in the pursuit of resources and the production of offspring.

The fact that intelligence enabled the evolution of reciprocity does not mean that we, or any other species of animal, actually calculate the probability that another animal will reciprocate, anymore than we calculate degrees of relatedness. Our altruistic responses are instinctual and automatic. This is also a product of evolution. In many cases an altruistic response must be immediate in order to be effective. If the animal were to consciously decide whether altruism was worth the effort and cost, the window of opportunity may have passed.

Reciprocity and Empathy

It is easy to imagine, as have hundreds of science fiction writers, intelligent species that lack altruism, such as the Klingons and Ferengi of *Star Trek*. Intelligence by itself may be a sufficient basis for the evolution of reciprocity, but it is not a very good one. As I mentioned earlier, the most successful behaviors are those that are reinforced by good feeling. Reciprocity works best and is most favored by natural selection when it is reinforced by feelings of empathy.

Empathy is often defined as the ability to feel what others feel. By this definition, empathy requires advanced intelligence. The empathetic animal must be able to reconstruct the mental state of another animal and experience it as its own. However, as primatologist Frans De Waal explains, the evolutionary roots of empathy go far deeper.[15]

First, consider that most mammals have a type of behavior known as emotional contagion. When we see another animal do something, we tend to imitate that behavior. For example, yawning, laughter, and applause are contagious. Emotional contagion has been demonstrated in experiments on mice, in which two mice were placed in separate clear plastic cylinders, close enough that they could see each other. When the experimenters annoyed one of the mice with acetic acid, which apparently gives the mice a mild stomachache, the other mouse would writhe as if in the same kind of mild pain. The other mouse would, in fact, become more sensitive to pain and react more strongly to being lightly touched with a hot object after seeing another mouse in discomfort than before. The mice displayed emotional contagion only when they could see another mouse, even if they could not hear or smell it. Mice in opaque plastic cylinders, in which they could not see other mice, did not react in this way.[16] De Waal points out that emotional contagion may be based on behavior that is even more ancient in evolutionary terms. Fishes and insects react to one another when they move, resulting in the astonishingly rapid reactions of entire schools of fish and in the massive movements of locust plagues.[17]

Second, mammals with greater intelligence often show empathy for other members of their species. In the mouse experiment described previously, the observer mouse reacted more strongly to seeing another mouse in pain, and was itself more sensitive to pain, if the observer and the victim had been raised in the same nest. The mice displayed empathy toward their littermates. Most mammals display empathy only sporadically, but apes (including chimpanzees and humans), cetaceans (such as whales and dolphins), and elephants are famous for their highly developed empathy. While the saying that elephants never forget is not literally true, elephants are good at remembering other elephants whom they have loved, and displaying signs of grief when reminded of them even years after they have died.[18] Apes, cetaceans, and elephants have special neurons (Von Economo neurons, or VEN cells) that may allow strong empathy. Humans whose VEN cells are damaged lack a well-developed sense of embarrassment, humor, self-awareness, and empathy. This type of neuron has evolved independently in the ape, elephant, and cetacean evolutionary lineages.[19]

Empathy can extend beyond species boundaries. We all know examples of animals that have formed what appear to be bonds of attachment with members of other species. Clergyman-naturalist Gilbert White published observations of interspecific empathy in the late eighteenth century. "There is a wonderful spirit of sociality in the brute creation," he wrote, "independent of sexual attachment." He proceeded to give examples, such as this description of empathy between a horse and a hen: "The fowl would approach the quadruped with notes of complacency, rubbing herself gently against his legs; while the horse would look down with satisfaction, and move with the greatest caution and circumspection, lest he should trample on his diminutive companion."[20] Most naturalists in Britain, including Charles Darwin, carefully read White's published observations.

Reciprocity and Education

Societies of intelligent, altruistic animals have also evolved systems of education. In such systems, the adults educate the youngsters in how to survive in the physical and social environment. This process has developed immensely in human societies, so much so that many evolutionary scientists consider education to be one of the reasons that humans can live for many years beyond their reproductive life span.[21] Old people are repositories of an immense amount of cultural knowledge, which they can impart to their offspring. Science writer Natalie Angier calls old people the "Alexandrian libraries of preliterate cultures."[22] What is most important in our present discussion is that educators teach not only their own offspring but also those of other members of the tribe. This is reciprocity: the educator instructs the students, and the other adults provide resources to the educators. This is the business I am in, and it is very satisfying. My noneducator friends like to say that it beats working for a living. But it *is* work, it has value, and I believe I am being self-interestedly altruistic by doing it.

Like many other characteristics of human society, education can be traced back to earlier species. In many animal species, youngsters learn by watching adults as they go through their daily activities. This is passive learning on the part of the young, but there is at least one nonhuman instance in which adults actively teach the younger animals. A real teacher is no mere role model for playful pups but rather undertakes deliberate actions that are intended to educate the young. Active teaching, as opposed to passive learning, has been demonstrated in meerkats (see figure 2-6 in chapter 2). Meerkats are members of the mongoose family of mammals, and as such are not human ancestors. They have evolved a system of education independently of humans. Meerkats live in African deserts and eat scorpions, which are dangerous prey for young meerkats to figure out how to catch and eat. They must be handled carefully, and the stinger must be disabled before the scorpion can be eaten. Student meerkats watch their teachers disable scorpions. Meerkat teachers, in response to hearing their cheeping noises, present student meerkats with scorpions. The teachers present the youngest students with dead scorpions but present their older students with progressively more scorpions that are alive but disabled, and then eventually with some intact and very dangerous scorpions. The adults alter their teaching style as the youngsters grow up, knowing the correct proportions of dead, disabled, and intact scorpions to present to the students based on the cheeps they hear: higher-pitched cheeps indicate young, inexperienced animals, while lower-pitched cheeps indicate older, more experienced animals. The researchers demonstrated this experimentally by watching adult teaching behavior in response to prerecorded cheeps played back by tape.[23]

Reciprocity and Trust

Reciprocity is greatly enhanced by trust. If one animal can trust another only as long as he can see him, reciprocity is limited in the benefits it can confer. But if one animal can trust another even behind his back, reciprocity can confer much greater benefits. In most animal species, this ability is quite limited. When the alpha male antelope or gorilla is not looking, the beta males can sneak around and find some action.

And no one will tattle on them, since they have no language! Memory and language allow a record of reciprocity, and of its violations, to be kept, and this is a dimension available only to humans. Once writing was invented, this process was even more greatly enhanced. Some of the oldest samples of writing are Mesopotamian clay tokens that appear to be records of commodities or of labor. Such records are more durable and reliable than spoken language.

Trust is never unconditional in adult animals. Very young animals have absolute trust, which is why young human children must be protected from adults who would take advantage of them or brainwash them. The only humans who are completely trusting beyond infancy are those who have mutations on chromosome number 7, which gives them Williams syndrome—a genetic condition that is characterized by some medical problems but also by qualities of friendliness, trust, and sociability.[24] Trust is a social shortcut: classifying others into categories of trustworthiness saves us the trouble of having to continuously assess the trustworthiness of each individual.

Human tribes in which the members cannot trust one another would quickly disintegrate when challenged by other tribes whose members can trust one another. That is, very altruistic tribes will triumph over less altruistic ones, and when a tribe is victorious, most of its members benefit—as do the genes that are within those individuals. Altruism creates a cascade of benefits: to tribe, individual, and gene.[25] Natural selection, therefore, favors reciprocity that is sincere and trustworthy—it also favors individuals whose minds take pleasure and experience inspiration for being trustworthy. Members of an altruistic tribe can focus all their energy on defeating another tribe, while in a less altruistic tribe, the members have to keep looking back over their shoulder for internal threats at the same time they are looking out for external threats. Only humans have the requisite intelligence to make a high level of reciprocity work.

Reciprocity and Fairness

In addition to empathy and trust, there is another emotion that reinforces reciprocity. It is a passion for fairness. Intelligent animals instinctively recognize when another member of their society has cheated and obtained benefits unfairly, and sometimes have a strong emotional reaction to it. Primatologists Sarah Brosnan and Frans De Waal found that capuchin monkeys reject food rewards when they observe other monkeys receiving a greater reward. These monkeys, at least when not starving, would rather receive no food than to tolerate an unfair distribution of it.[26] Other recent research demonstrates that dogs recognize and resent unfair rewards.[27]

The emotion that reinforces fairness is sometimes called "sweet revenge" and sometimes goes by its German name *schadenfreude*. Intelligent animals derive pleasure from seeing the humiliation, and sometimes even the suffering, of other animals who have gained their advantages unfairly. This is well known in humans, and I will discuss it again later. Schadenfreude has also been observed in chimpanzees. On one occasion, two male chimps who observed another male chimp receiving special treats from human visitors escaped from their enclosure and attacked the man who brought the treats. The chimps chewed off most of the man's face and buttocks, ripped off his foot, and bit off both of his testicles, before they were shot.[28]

Intelligent animals also fear sweet revenge by other animals. Bonobos (the more peaceful relatives of chimpanzees) have been observed to reject food when other bonobos without food are watching them.[29]

Direct and Indirect Reciprocity

And there is yet more to the story of human reciprocity. We have so far examined *direct* reciprocity—in which a meerkat or monkey or chimp or person carries out an unselfish act with the expectation, whether instinctual or calculated, that the recipient of the generosity will later return the favor. But there is another kind of reciprocity that may be found only in the human species. It is *indirect* reciprocity, in which a person carries out an unselfish act for the benefit of a poor or weak member of society who will probably never have the opportunity to return the favor. You may be rewarded if you are nice to a powerful member of your tribe—but what possible reward can you get for being nice to someone who is down and out?

The reward for indirect reciprocity is reputation. A person who is generous to fellow humans who have experienced misfortune demonstrates that he or she is rich enough that he or she can afford to give some of it away. But there is more. Since altruism and empathy are so deeply engrained in the human mind, most of us will genuinely admire a generous person without even thinking about his or her financial status. A generous man may obtain more reproductive opportunities as women look at him and admire his generous character. A generous woman would likewise be admired by men. Anyone would be more willing to trust a truly generous person than a person who is known to use generosity only as a way to advance his or her self-interest. In indirect reciprocity, therefore, the altruist expects a payback not from the recipient but from the onlookers. As any reasonably intelligent person will tell you, a good reputation is more valuable than a lot of money in the bank. Nice guys do not necessarily finish last, because they may not have to run a race in the first place.

And they do, in fact, usually expect some sort of payback. Rich people typically expect to have buildings or scholarships or programs named after them. Consider the story of the rich banker who donated so much money to a certain university that they named not just a building but an entire school after him. He undoubtedly received at least some payback for this because his bank got more customers, who felt that they could trust their savings and mortgages to so generous a man. It pays to advertise—and conspicuous generosity is one of the best advertisements. The banker's reputation for public service has also gained him an incalculable level of influence on the higher education system of the state in which he lives, including the ability to nearly handpick university presidents, sometimes with disastrous consequences. Conspicuous consumption is one way of creating a reputation—many people could buy an entire house for less money than this banker spent on his front door (just the door)—but conspicuous altruism is an even better way of creating a reputation.

Conspicuous altruism can also clean up a tainted reputation. Consider the story of a very rich designer of computer software who gained a virtual monopoly on operating systems. His reputation was not good—he was known to be a difficult taskmaster and ruthless in the suppression of competitors. Any of us who use the software marketed by the company he started can testify that it has numerous and maddening glitches. All publishers with whom I have worked use it, and I have no choice but to put up with these glitches, such as the aptly named cursors jumping to random spots on the page as I am writing these words (there it goes again!). But once this man was comfortably rich (he earns in a few seconds what I earn in a year), he became a great philanthropist. The foundation started by him and his wife is without question a significant force for good in the world, and they enjoy universal admiration for it.

So passionate is the human love of altruism that one of the most despised characters in all literature is the man who rejected all three kinds of altruism: Ebenezer Scrooge. He was mean to his nephew, who was the only way his genes could ever get into the next generation. He was mean to his employee Bob Cratchett, and he was mean to the people who came around asking for a donation to help the poor. No kin selection, no reciprocity. Not surprisingly, he was single and had no children. Dickens portrayed him as a failure of the human spirit, but Scrooge was obviously an evolutionary failure as well. He failed to act in his own self-interest by tapping into the benefits—and joys—of being an altruist.

Another component of the complex human capacity for altruism is something you probably think belongs in the previous chapter: homosexuality.

Sexual behaviors are primarily the result of sexual selection because they enhance reproductive success. This is where they came from. But once sexual behaviors evolved, together with the hormones and the intense emotions, they became available for other uses. Sexual interactions between members of the same gender are widespread among animals. Sometimes they are used to establish dominance, especially by one male over another. Perhaps most often, sexual pleasure is used as a means of reinforcing bonds of altruism between individual animals. Just as sexual selection favored the evolution of heterosexual behavior in the first place, natural selection subsequently favored the evolution of homosexual behavior within complex animal societies. Sex reinforced altruism.[30] There apparently is a genetic basis for male homosexuality; gays respond to the scent of other males the same way most women do. Female sexuality seems to be more flexible than that of males. The sexual feelings, whether favored by reproductive success or by altruism, are the same.

Unselfish Altruism

Once in a while we meet or hear about truly unselfish people who sacrifice themselves for the good of others without the possibility of any reward whatsoever, even indirect. Many people would consider Jesus of Nazareth to have been such a person. He made it very clear that being good to other people with the anticipation of gain was not worthy of godliness. In his famous Sermon on the Mount, he asked, "If you salute only your brethren, what more are you doing than others? Do not even the Gentiles do the same?" He thus dismissed what we today call kin selection. "If you love only those who love you, what reward have you? Do not even tax collectors do the same?" He thus dismissed what we today call direct reciprocity.

> Beware of practicing piety before men in order to be seen by them.... When you give alms, sound no trumpet before you, as the hypocrites do in the synagogues and in the streets, that they may be praised by men. Truly I say to you, they have their reward. But when you give alms, do not let your left hand know what your right hand is doing, so that your alms may be in secret. ... And when you pray, you must not be like the hypocrites; for they love to stand and pray in the synagogues and at the street corners, that they may be seen by men.

He thus dismissed what we today call indirect reciprocity. The point I am trying to make in this chapter is that inclusive fitness and reciprocity are the evolutionary rewards that have caused altruism to evolve within the human species. Now that these behavior patterns exist, now that our brains have the capacity for them, they can be applied unselfishly, although few people do so. Jesus, Francis of Assisi, Mother Teresa, and Mahatma Gandhi are memorable examples. Chico Mendes and Sister Dorothy Stang were shot to death while defending the rights of peasants to stay on rainforest land that was being illegally logged in Brazil.

Throughout history, soldiers who conquered enemy cities or nations had many reproductive opportunities with the vanquished females, and the ones who returned home were heroes, with yet further reproductive opportunities. But most soldiers believe passionately that they are defending their country and their families. The soldiers on the first wave of the D-Day invasion were engaged in a virtual altruistic suicide mission. The soldiers, of course, had no choice in the matter, other than to face shame and imprisonment. But were it not for the passion of altruism, a lot of them would undoubtedly have chosen shame and imprisonment over death. And the soldiers came to think of their fellow soldiers as family members, a band of brothers (and now sisters). They thus added the emotional underpinnings of kin selection to the social bonds of reciprocity.

Not only are most soldiers motivated primarily by altruism toward their fellow countrymen, but most soldiers also lack a killer instinct. Even during World War II, only one soldier in five actually shot enemy soldiers.[31] Even terrorists usually have feelings of empathy until those feelings are systematically programmed out of them by their leaders. Terrorist training typically includes desensitization, in which recruits watch videos, over and over, of victims being beheaded.[32]

Soldiers face death even if they have doubts about the truthfulness and sincerity of their leaders—even if they suspect that the wars in Vietnam and Iraq were started by Democratic and Republican politicians based on lies and for the purpose of obtaining lucrative taxpayer-funded contracts. It was the sincere altruism of the soldiers, contrasting sharply with the greed of the politicians, that turned many conservative Americans against President George W. Bush and Secretary of Defense Donald Rumsfeld—especially when wounded soldiers returned to their country and were greeted by cockroaches at Walter Reed Army Medical Center. Abraham Lincoln wrote a deeply emotional letter, with his own hand, to a mother who had lost five sons in the Civil War. Could anything contrast more greatly with this than Donald Rumsfeld's machine-signed letters of condolence to the families of slain soldiers in Iraq and Afghanistan?

THE DEATH OF ALTRUISM

So there it is—altruism, one of the most amazing phenomena ever seen on Planet Earth, and one that has changed the Earth itself. Human altruism evolved mostly during the 90 percent of human history that was spent as hunter-gatherer tribes. In such tribal situations, everyone knew everyone else, and nobody could hide from the consequences of their violation of altruism. There was no place for liars and cheaters to go, except into banishment, perhaps to be taken in by another tribe, perhaps not. In some cases, as with the Cherokees, the banishment was temporary, and the life of the entire village, including its social life, was renewed every spring when they burned their old clothes and their old food and swept the streets.[33] Clearly, there were abuses of power by chiefs in prehistoric villages. But a leader could afflict a fellow tribesman only after looking him in the eye.

Civilization and Altruism

But altruism is dying. It has been dying for about five thousand years, and its morbidity is accelerating today.

It began with the evolution of agriculture. In prehistoric times, as the most recent ice age was ending, an abundance of wild grains allowed a large population—large at least by hunter-gatherer standards—to subsist on wild foods. Over the centuries, the gatherers selected and encouraged the growth of their favorite wild plants. Perhaps as a result of the return of a more challenging climate, they invented agriculture—but only after a long period in which the characteristics suitable for farming had evolved in these plants.[34] This happened independently in several parts of the world, such as the Middle East, China, Africa, Mexico, and the Andes.

Agriculture was a smashing success in terms of food production. It only took a few people to raise enough food for everybody. But it also caused nomads to settle down and allowed armies of professional soldiers to be fed. The result was the world's first city-states. People captured from vanquished villages could be made into slaves to raise the food. City-states became empires. They were large enough, and their dictators sufficiently powerful, that they did not need to earn any respect from the people. All they needed was the support of the armies. It was now safe for them to substitute force for altruism. This has occurred in an evolutionary blink of the eye. The human species still has altruistic brains, but its leaders parasitize altruism for their own power and fortune.

Altruism in Recent History

In recent decades, altruism appears to have become more common. Even as recently as World War II, people of civilized countries thought there was absolutely nothing wrong with killing thousands of civilians who happened to live in an enemy country. The firestorm of Dresden and the atomic bombs dropped on Hiroshima and Nagasaki seemed reasonable to Americans, even though very few of the victims were actually responsible for German and Japanese aggression. The Rape of Nanking and the conquest of Europe seemed reasonable to Japanese and German citizens. While, as mentioned earlier, soldiers had a hard time shooting fellow human beings, most soldiers and civilians approved of mass bombings of civilians whom they did not have to look in the eye. Somewhere around the time of the Vietnam War, this attitude changed. It was no longer acceptable to massacre a village, such as My Lai, just because there might be some enemy combatants there. Today, whenever an American bomb kills civilians in Afghanistan, there is a worldwide uproar. All around the world, people of every religious conviction or of no religious conviction are uniting in their rejection of torture, genocide, and war-related cruelty. This sounds like good news. I cling to it, because it is almost the only good news about the direction the world is headed.

But we must remember that this altruistic progress is the result of the beliefs and actions of individuals rather than of governments. Governments, at best, acknowledge the human rights that their people demand, and at worst, suppress them. Government administrations do not advance human rights. The American government responded to civil rights leaders like Martin Luther King Jr. first with hostility, then with acquiescence, and only after many years with admiration. The progress of altruism has always and only come from the bottom up. When altruists find themselves in positions of power, they also find themselves in positions of frustration and are seldom able to accomplish very much.

And it is usually not facts and figures that stir people's hearts to create a change. It was not the list of deaths and battles in Vietnam that altered American opinion; it may have been a single Associated Press photograph of children running from the village of Trangbang on June 8, 1972, screaming in pain from the burning napalm with which they had just been doused. We are still an altruistic species, and when we see something like that, it moves the hearts of everyone—once again, with the exception of psychopaths.

The Suppression of Altruism

Just as many governments abuse their citizens, administrators abuse their public and private employees. Administrations of governments, universities, and corporations have institutionalized the repression of altruism to a breathtaking extent. Very commonly, they are so hostile toward altruism that they make decisions that cost them dearly in profits and reputation. They create a fortress of secrecy within which the leaders demand absolute loyalty. Administrators restrict access, so that only one or a few persons can address them. It is not uncommon for altruistic employees, as soon as they have a good idea, to be swiftly suppressed by administrators, as in a giant game of Whac-a-Mole. The customer service representative on the other end of the phone line may sincerely want to take your side in a dispute, but just let her try! The climate of intimidation and secrecy was one of the main reasons that illegal activities progressed within Enron to the point that it collapsed.

One result of the suppression of altruism in both public and private sectors is that self-correction becomes impossible. The illegal financial activities of Enron, and the deceptive economic practices of the top investment banks in the United States, were revealed only when they became so large that they could no longer be hidden. The worldwide financial "meltdown" began in the United States, and the administration of President George W. Bush and major financial corporations are largely responsible for it. The current financial crisis has demonstrated, not for the first time, that corporate selfishness is a recipe for not only moral but financial chaos. President Franklin Delano Roosevelt said in 1937, "We have always known that heedless self-interest was bad morals; we know now that it is bad economics."

It is not the desire for profits that has caused the centuries-long decline in altruism and the current financial crisis. Every business that has ever existed has tried to make as much profit as possible. That's what business is about. But when businesses existed within communities, they could not get away with dishonest practices, or if they did, their collapse was localized. But now that community businesses have become global corporations, dishonest and destructive practices can spread along the tendrils of administration and management and infect their offices all over the world. Rather than being accountable to their neighbors, they are accountable to no one but themselves. When this happens, we reward them. They declare themselves to be "TBTF," or too big to fail; they demand, and receive, billions of dollars of government aid.

Meanwhile, the CEOs of the top financial corporations in the United States seem to be utterly oblivious to the schadenfreude resentment of millions of Americans. Many of these executives received millions of dollars of compensation even while their corporations were losing money and receiving government assistance.[35] When the leaders of the top three American automakers first came to Washington, DC, to ask for government assistance, they flew in corporate jets. They were surprised that congressional leaders reacted in anger. These executives then drove to Washington in the smallest cars their companies manufactured (according to one cartoonist, these were the Ford Implorer, the Chevrolet Grovel, and the Chrysler Mendicant). Some CEOs paid back small amounts of their compensation, but their corporations quickly went back to paying them huge bonuses. Americans are angry. Their anger does not come from envy. Instead it comes from resentment of the unfairness of large compensations given to CEOs as a reward for failure.[36] I wonder how many Americans would like to do to them what those two male chimps did to the unfair intruder.

President Barack Obama was swept into office on a tide of enthusiasm, but he is one idealistic altruist, allied with a few other idealistic people, and he can only work through a slow and contaminated mass of administrators. At least President Obama has made efforts to create a "transparent" administration, even to the extent of placing minutes of departmental meetings online. This could not contrast more strongly with the preceding administration, particularly the notoriously secretive office of former vice president Dick Cheney. We Americans dare to expect our government to be altruistic. The majority of Americans believe Barack Obama is altruistic, but we remain unconvinced about his appointees, many of whom were financial executives receiving large compensations.[37]

Altruism evolved in communities. The death of communities, and their replacement by multinational corporations, may prove to be the death of altruism. When, for example, all food is produced by a few global corporations, we have to take what they offer. Contrast this with farmers' markets, where you buy food face-to-face from the people who produce it. When the people who run a community or a state live in it, they are much more accountable to their neighbors. This is the reason that conferences of mayors or governors have much more bipartisan cooperation than Congress does. The local altruistic citizenry of nearly every community wants energy efficiency and a clean environment, but Congress listens to the moneyed interests of the coal, oil, and gas industries, who want us to burn as much fossil fuel as possible. When large corporations and political parties are in control of the world, there is no voice for altruism.

I heard a song on the radio, whose title (and only lyric, as far as I could tell) was "I don't care about nothin' but me." A person who lives in this manner has forfeited all the evolutionary benefits of altruism. There have always been people who believed this, but today corporate media hypnotize many more people into believing it. Corporations benefit by having people buy things to gratify themselves, not by having people help one another.

The Future of Altruism

We have already gotten a few glimpses into what the world could be like if many countries suffered a breakdown of altruism. We call them "failed states." The genocides in Rwanda and Bosnia in the last decade of the twentieth century, or in the Sudan in the first decade of the twenty-first, show how people who had been living with a semblance of peace can suddenly erupt into insane violence. Just as the human brain is capable of altruism, it is also capable of classifying other human beings as nonpersons and killing them with no feelings of empathy whatever. This is unlikely to occur except under highly unusual circumstances, but it is clearly possible.

It would take thousands of years for the genetic underpinnings of human altruism to erode away. Unfortunately, the cultural norms of altruism are essential for unlocking these genes. If we interrupt the cultural transmission of altruism memes, we may collapse into a nightmare world of dark conflict in which the old altruistic tendencies are groping around, unable to emerge. It would take thousands of years for meerkats to lose the ability to digest scorpions, but it would take only a generation, during which the meerkat education system is disrupted, for them to forget how to catch them efficiently. The interruption of the cultural transmission of altruism is now possible on a worldwide scale, since there is literally no community on Earth that cannot be directly impacted by the military and economic might of the world's dominant countries. The human economy is so large that the collapse of altruism could bring about a disruption of Earth's life processes rivaling that of the Permian extinction or the asteroid that ended the Cretaceous period.

This is Earth's midlife crisis. Earth cannot die, but it can be battered. Earth's human children have grown into monstrous but still immature adolescents who, unlike the mindless Permian clouds of hydrogen sulfide and the Cretaceous asteroid, actually know how to destroy the planet. The only way to save the life support systems of Planet Earth is to save altruism. No government or program can do this; only the united voices of billions of altruists, unafraid of actions that administrations, management, and dictators may take against them, can do it.

In the Judeo-Christian story of the Garden of Eden, God asked Cain about his brother Abel, whom he had just killed. Cain had a famously smart-assed response: "Am I my brother's keeper?" And today, many people who deny their own innate altruistic feelings look at environmental destruction and think, "So why should I care?" At the moment, environmental problems such as rising ocean levels and droughts are affecting only the poor and powerless. At the moment, Americans are rich enough to buy our way out of the environmental effects of global warming and pollution and the depletion of water resources. But we need to reaffirm altruism. We are responsible for our biological families and our metaphorical brothers. I am my brother's caretaker and it is my brothers' and sisters' Earth.

Chapter 7

RELIGION

There is no such thing as religion.

Many writers assume that religion is a unified set of evolved behaviors, as I did in the original edition of *Encyclopedia of Evolution*. (I have corrected this mistake in my revised encyclopedia.) It is now clear to me that religion is not a single thing; it is a set of memes (see chapter 2) that have taken up residence in human minds. These memes use human minds, words, and actions as a way of propagating themselves. The human brain is the hardware, and religious memes the software. We have reified these memes and the physical attributes they use into a single concept. I hereinafter use the term *religion* to refer to this collection of memes.

Therefore when we say that religion is universal among humans, we mean that every human has the mental components that can or do harbor and propagate religious memes, and that some of those memes can be found in every culture and every individual. We cannot say that natural selection has or has not favored religion as a whole. Natural and sexual selection clearly controlled the origin of the brain processes of which the religion memes make use. And social evolution has promoted religion in most cultures at most times, often to the benefit of powerful individuals who use it to dominate others. But we cannot say, as I once said, that religion per se is an ineradicable part of the human mind.

This chapter about religion might seem to be a digression from the story of evolution, but religion is a unique genetic-memetic adaptation of the human species and has been one of the most important reasons that humans have had a powerful impact on the Earth. This chapter is about the human phenomenon of religion, not about whether religion does or does not have a separate reality beyond the human mind.

THE BRAIN AND RELIGION

Only humans appear to have religion. Intelligent nonhuman animals appear to live only in the present. They seem unaware that they will die, nor do they take much notice when other members of their species die. The human awareness of death, an awareness we carry with us from adolescence onward, is not necessarily superior to the blessed ignorance of most nonhuman animals.

Exceptions may include elephants, who appear to mourn and to remember the deaths of specific individuals, and chimpanzees, our closest living relatives. A vivid example appeared in the November 2009 *National Geographic*. The chimpanzees at the Sanaga-Yong Chimpanzee Rescue Center in Cameroon all knew one another. And they all knew an old female, who died in her late forties, which is an advanced age for a chimp. As the human caretakers carried her away on a stretcher, the other chimpanzees lined up along a fence and did things that, to a human observer, appeared to be expressions of grief. It was certainly empathy (chapter 6). Several researchers, including anthropologist Jane Goodall, have observed chimpanzees appearing to dance as they watched a waterfall, as if they were exulting in its beauty. While even Jane Goodall cannot know what was going on inside of the minds of the chimps, her interpretation—that this was proto-religious awareness on the part of the chimps—is credible.[1]

The main reason that fully developed religion is unique to humans is that humans have such large brains. The size of the human brain relative to the size of the body, compared to that of other primates, is out of the ballpark. Relative brain sizes of primates exceed that of the average mammal, which exceeds that of the average vertebrate. More than any other species, humans seem to have excess brain capacity. One of its outlets is religion.

This would not necessarily be the case in other intelligent species. Science fiction has created a universe full of intelligent but essentially nonreligious species. For the inhabitants of Middle Earth, created by the Christian writer J. R. R. Tolkien, religion is not distinct from everyday life. Intelligence does not require religion, but in our case the increase in brain size over evolutionary time has resulted in the capacity for religion.

Religion over Evolutionary Time

Brain growth began with the evolution of the genus *Homo: Homo habilis*, then *Homo ergaster*, in Africa. There is no evidence that they had religion. Some populations of *Homo ergaster* migrated out of Africa to Asia, becoming *Homo erectus*, and to Europe, becoming Neandertals (*Homo neanderthalensis*). Some remained in Africa, where they evolved into modern humans (*Homo sapiens*). There is no evidence that they had religion either. How, you might ask, would we know? One indication is that they did not create any artwork that suggested religious sensibility. In fact, they did not create any artwork, period. This, despite the fact that some *Homo erectus* and all Neandertals had brains about as large as those of modern humans. Some African humans migrated to other places, and some stayed behind, resulting in the evolution of the modern human races. Religion eventually evolved in all those human races, perhaps independently.

The most dramatic evolution of what appears to be religion was in the modern humans who migrated to Europe, often called the Cro-Magnon people. The reason is not clear, but it may be that only in Europe did *Homo sapiens* encounter other human species that had comparable levels of intelligence. African *Homo sapiens* were clearly more intelligent than the remaining *Homo ergaster*; Asian *Homo sapiens* were clearly more intelligent than the *Homo erectus* they doubtless encountered. (They probably also encountered the little Flores Island "hobbit" people, a separate species called *Homo floresiensis*.) But when the Cro-Magnon encountered the disturbingly intelligent Neandertals, there was a challenge: no complacent level of intelligence was enough for them. The encounter between *Homo sapiens* and Neandertals may be considered a natural experiment that compared an intelligent religious species to an intelligent nonreligious species. In the struggle for existence, the Cro-Magnon developed spectacular art that is usually interpreted as religious. Religion, however vaguely defined, may have given the Cro-Magnon the social cohesion that helped them prevail over the Neandertals, whom they gradually drove to marginal habitats where they died out.[2]

Hardware: "Religious Parts" of the Brain

Human brains increased in size for a number of reasons. As explained in chapters 3, 5, and 6, the mastery of technology, sexual selection, and social interactions were the major reasons for this increase; religion probably had nothing at all to do with it. But as the brain increased in size, it was not just the parts of the brain that conferred social and technological skills that increased; the whole brain increased in size. Some parts of the brain, even while not themselves conferring advantages, may have gotten dragged along in the general brain size increase. Here are some examples.

- *Sexual ecstasy*. Humans have a highly developed capacity to experience sexual ecstasy. Religion, like orgasm, can make you feel that you have transcended out of your place on Earth (*ex-* means "out of," *stasis* means "place"). Religions connect religious devotion to sex, for example, in the overtly sexual imagery of the Song of Solomon or the poetry of Hildegard of Bingen.
- *Loss of the awareness of having a defined body*. Nearly everybody takes the awareness of his or her body for granted. Once in a while, a person experiences a certain kind of stroke in which parts of the left hemisphere are deprived of oxygen and fail to work while the right hemisphere continues to function. Neurobiologist Jill Bolte Taylor describes a stroke in which she had a feeling that she was not confined to her body, and that she was flowing through a stream of experience.[3] Our brains take nerve impulses from every sensory organ—and these impulses are all the same—and sort them out into sound, color, pain, heat, cold, taste, smell, and orientation. That is, we detect light with our eyes, but we see with our brains; we detect airborne molecules with our noses, but we smell with our brains. Sight and smell are illusions created by the brain. Apparently, the sense of being confined within a body is also something the brain creates. In ancient times, some people may have experienced head traumas, oxygen deprivation, starvation, or dehydration, which opened them up to a disembodied sensation, or they may have induced these feelings by meditation. This is one of the elements of religious experience commonly reported by people who have a well-developed ability to meditate.[4]
- *Altruism*. Altruism (chapter 6) is one of the most pervasive human characteristics, one for which religion clearly provides an outlet.
- *The need for an authority figure*. Humans appear to have a psychological need for an authority figure whose goodness they do not question. In adults, this is a neotenous vestige of a child's worship of parents.

People continue to follow their charismatic religious leaders long after the leaders' hypocrisies are revealed, and break themselves from this bond of devotion only after a great deal of anguish.

- *Awareness of death.* Natural selection favored the evolution of intelligence, and one of the side effects of intelligence is the ability to understand that you will die and the possibility that you will be preoccupied by it.
- *Agency.* Very young children do not display feelings or awareness that can easily be described as religion. They do, however, always have the capacity for agency attribution. When something happens, they think someone has caused it to happen. The wind blows because someone makes the wind blow. If they experience pain, it is because someone is hurting them. The agent of wind or pain is invisible, but the children believe in the agents anyway.[5]

Religion makes use of all these brain elements.

The Tunnel

And then, strangest of all, there is the near-death experience. The simplest form of near-death experience is one in which people see their entire life story rerun in front of their eyes when they are mortally threatened. This has been reported so many times that it has become a cliché. In one of his comedy routines, Bill Cosby said that, as a little kid, he had such an experience, but his life had been so short that the story had to run twice.

When a person draws near to death, she reports that she is looking down on her own body; he feels bliss; she meets and talks with loved ones who have died; he proceeds through a tunnel toward light. The only way we know about this is that many people who have come close to death, or actually experienced clinical death, have been revived and have told us about their "near-death experiences." They uniformly report that they did not want to return to the land of the living.

Most of the people who have had these experiences believe strongly that they were actually looking into heaven, and that a spiritual realm exists that is so real and important that daily human life is inconsequential in comparison. Many of these people become quite inconvenient and annoying to those of us who have not had their experience, since they seem unconcerned about taking care of the normal responsibilities of mundane life, which, to them, is trivial compared to eternity. Whether or not they are Christians, they are just about the only people who really, truly believe what Jesus said about not giving a thought for the morrow.

Many of you have probably had the same experiences I had. As I learned about evolution, I had to give up creationism. As I read the Bible and recognized its human authorship, I had to give up simple biblical faith. But it was not easy. I wanted to cling to something. The reports of near-death experiences gave me something I thought I could cling to. Were these people *really* seeing into Heaven? If so, then there really was a Heaven. I did not need to base such a belief on questionable biblical passages; I could base it on the eyewitness accounts of people alive today. But how do you know they were not hallucinating? Oxygen deprivation could make their brains create all these sensations.

There seemed to me to be a couple of reasons to believe that these people were not hallucinating. First, the repertoire of near-death visions seemed remarkably uniform: seeing their bodies, seeing loved ones, and the tunnel. This does not appear to be a generalized hallucinatory experience. Something this specific would have to be the product of natural selection, and it is difficult to think of a way that natural selection could have produced such a specific a set of hallucinations. Of course, it is not certain that these experiences are, in fact, as universal as they seem. The patients whose accounts are most widely read are Westerners, and they may have interpreted their experiences in terms of what Western culture considers a vision of heaven to be.

Second, some of these people report having seen things they could not have known about. They report events that occurred in the operating room while they were in a coma. Of course, it is by no means certain that a person who is considered clinically comatose is totally disconnected from gathering sensory experiences for later recall. But one person reported seeing a pair of red shoes on the roof, and, as the story goes, this turned out to be true.

Wow. These bits of information are tantalizing. What we need is a thorough investigation of near-death experiences.

Enter Sam Parnia, MD, who wanted to use an experimental approach to study near-death experiences. If the person's "spirit" actually hovered above the body, he reasoned, then it should be able to see symbols printed in black-and-white on the *upper* side of ceiling tiles. Parnia arranged for some of these tiles to be installed in emergency and operating rooms, and made sure that the doctors and nurses did not know they were there—otherwise, the medical personnel could subconsciously influence the patient's knowledge of them. I read Parnia's book with a growing sense of excitement.[6] What were his findings?

Nothing. His book merely described the experimental setup. Apparently, he is saving his results, if any, for future publication. We are waiting, Sam. Meanwhile, we must proceed with the assumption that near-death experiences are hallucinations.

There are, in fact, good reasons to believe that these visions *are* hallucinations. The tunnel experience can be induced artificially by electromagnetic stimulation of the right temporal lobe and, not quite as effectively, by certain drugs, such as ketamine.[7]

The point I am making is that one of the consequences of the enlargement of the human brain was that the right temporal lobe became capable of producing vivid hallucinations. Prehistoric people who appeared to have died, then recovered, would report what they thought they had actually seen, providing what seemed to be a factual basis for religious beliefs.

Software: The Memes of Religion

The human brain has a desire to understand and explain things. This is the major component of evolved human intelligence. If you have any of the brain experiences mentioned above—a sense of disembodiment, a conviction that everything is caused by an agent, sexual euphoria, or visions of the tunnel—you will also have the compulsion to explain them. Here is an empty space just waiting for memes to walk in. One meme told people that there must be something beyond death. Another meme told them that there must be spiritual beings causing everything from the wind to the rising and setting of the sun. Another meme told them that their experiences of disembodiment and the tunnel were actual observations of a spiritual realm. Put these together, and you have a primordial religion. The only alternative, to a prehistoric person, is to ignore the whole thing, but the most successful people were those who figured things out, not those who ignored them. Even if the resulting beliefs were incorrect, so long as they enhanced the survival and reproduction of the believers, natural and sexual selection would favor them.

Human creativity is irrepressible, and it was inevitable that humans would couch their explanations of spiritual experiences in terms of mythological stories that addressed each of the brain phenomena, and that they would develop practices that enhanced the experiences. The stories had gods who made things happen. Religious practices were sometimes accompanied by hallucinogenic drugs, which tapped into the sense of disembodiment. Religion fed on sexual feelings in two ways: adherents experienced sexual feelings both about their deities and about charismatic religious leaders. Religion made us eager to follow charismatic leaders who claimed to have a connection with the gods. We humans had a desire to understand these overwhelming experiences—what young person has not tried to express the overwhelming sense of being in love?—and religious memes satisfied this desire.

The earliest religions may have been like the theology of Adam and Eve, as depicted in the early chapters of Genesis. They seemed to have no theology except that they knew there was a godlike being who walked around in the garden in the cool of the evening. Deep in the Cro-Magnon caves, such as Lascaux and Altamira and Chauvet, people seeking religious experiences made lots of marks on the walls and touched the walls, painting silhouettes of their hands. According to anthropologist David Lewis-Williams, they thought they were making contact with the spirit realm.[8] It may have taken thousands of more years before complex mythologies and systems of theology developed.

THE POWER OF RELIGIOUS MEMES

Religious memes are among the most powerful that have ever conquered the human mind and then used it as a vehicle of propagation. They may be the most powerful mediators of human experience and passion. A list of such examples would be as long as human history. I will just offer three.

Fruit Punch and the Planet Clarion

In the 1950s, a Chicago woman, Dorothy Martin, claimed to have received messages by automatic writing from her mentor, Sananda, on the Planet Clarion. She convinced a number of followers that the inhabitants of Clarion would destroy most of North America, starting with Chicago, on December 21, 1954. These followers quit their jobs, sold their possessions, and came to her house on December 20 to await the spaceship that would save the faithful followers. As midnight passed, the followers wondered what was going on. At 4:45 in the morning, Martin claimed to have received another message, in which Sananda announced that the attack had been called off. Martin ended up in Arizona, where she claimed (under the name of Sister Thedra) to receive messages from Sananda until her death in 1992. One person, on the basis of flimsy evidence, had convinced other people of something ridiculous (the name Clarion should have been a dead giveaway); this indicates that the human brain is primed to look for and cling to memes that claim to be religious revelations.[9]

Marshall Applewhite was a director, opera singer, and music professor who had psychological problems. In the 1970s, he gathered followers into his Heaven's Gate cult, convincing them that he was the reincarnation of Jesus Christ and that salvation would come in the form of a spaceship. In 1997, he told them that the spaceship had come and was in the tail of the comet Hale-Bopp. The whole cult committed suicide, by poison, alcohol, and asphyxiation, in order to board the spaceship.[10]

A preacher named Jim Jones convinced hundreds of people in the United States that he was the supreme representative of God in the modern world. He began to fear that government authorities were prying into his church, the Peoples Temple, and that they might discover something incriminating or curtail some of his activities. So he and his followers moved to Guyana and established a village in the jungle called Jonestown. Jones still expected that the end of the world would come in the form of a government takeover of his church, so he had his followers practice mass suicide by drinking fruit punch. A 1978 visit from US congressman Leo Ryan seemed, to Jones, to be the fulfillment of his fears. Cult members met the congressman at the airport and gunned him down. Jones then told his followers to drink the cyanide-laced fruit punch: this was not a practice drill. They all did it, even Jones himself. Acres were covered with the colorfully clad corpses of the suicide victims.[11]

This is how powerful religion memes can be. Jim Jones had created a mythology about himself, which incorporated many of the evolved capacities of the human mind. His followers thought of the Jones cult as their family, which tapped into their desire for altruism. He made use of the power of sexuality, convincing his female followers that he was not only their spiritual leader but their physical husband, as well. Adherents gave up everything to join his cult, and when it was threatened, they believed Jones when he told them they had nothing more for which to live.

My Experiences with Religious Memes

Religion memes can take over a mind to the point that it supplants most other human thoughts and connections. The example I know best involves my own grandfather. Neither I nor any other living descendants of Jacob Aville Rice know much about his early life. Jake and his companions would travel from one Holy Roller church to another and preach. There is no evidence (from any subsequent family wealth) that money was the motivation. Jake devoted himself so much to his religion that he left his family without enough money to meet their essential needs. He virtually ignored my father, who nearly died in childhood from an infection that followed appendicitis. My father carried the physical scar of the surgery, and the mental scar of his father's emotional rejection, for life.

Jake had read in the Bible that in the latter days, which he assumed to mean the 1960s, children would hate their parents, and parents would hate their children. He understood this to be prescriptive, not descriptive, and he felt obligated by his religion to hate his children. Old home movies show that he was not entirely successful, as he smiled and joked with them a lot when they were adults. His voice on a 1943 homemade record, sent to my parents, did not sound hateful. And the Bible said nothing about hating grandchildren, so scripture permitted him to be very nice to me.

There may be three possible responses that children can make to a parent who sacrifices their emotional lives on the altar of religion. One response is to practice a nurturing and constructive form of religion. Two of my uncles did this. Another response is to reject religious practice. This is what my father and aunt did. I discuss the third response in the next section.

My own experience of religious conversion to fundamentalist Christianity was probably the most powerful emotion I have ever felt, more powerful even than being in love. I felt like I was swimming in an ocean of joy, everything seemed to glow with invisible color, and the very air of the San Joaquin valley was sweet, no matter how much dust, pesticide, and exhaust fumes it contained. No wonder that, for centuries, many Christians have interpreted this feeling as actual inspiration by the Holy Ghost.

I believe that I have demonstrated my point, that religion can be a very powerful force, the most powerful force, even the exclusive force, in the human mind. How can one describe a person whose entire reality is religion, except that the person is to some degree crazy? Certainly in the case of many of the famous religious figures from history, it was difficult to distinguish mental illness from religious ecstasy. Centuries ago, there was no recognition of mental illness; religious passion was considered to be either from God or from the devil.

RELIGIOUS PARASITISM

Once religions and mythologies were in place, they became a resource whereby one person or group could manipulate another person or group. That is, they became the medium through which religion memes and religious leaders could become parasitic.

Richard Dawkins explains some of the characteristics that make religious memes so effective. Humans have a desire for some kind of faith; many religion memes offer false nutrition for this desire. But that is not all. As Dawkins explains, very religious people "make a positive virtue of faith's being strong and unshakable, *in spite of* not being based upon evidence. Indeed, they may feel that the less evidence there is, the more virtuous the belief" (emphasis in original).[12] A related meme is one that says that the article of faith is a "mystery" that the human mind will never be able to understand, and the fact that the article of faith seems like a howling self-contradiction is actually evidence of its truth. An example is the meme that says God is Love but that God will send everyone who is not a member of your church into hellfire, where they will suffer consciously and eternally. True believers will not even notice the contradiction between a God who is infinite love yet who will torture somebody forever without having given them a chance to avoid their fate. Some Christians even believe, as the fourth-century cleric Tertullian of Carthage taught, that the blessed spirits in heaven will be able to watch the torture of the damned in hell forever.

To the person whose mind religious memes have subverted, any evidence against the memes proves them to be right, not wrong. Weaknesses become strengths. This is a positive feedback system that makes the believer fall ever deeper into religious memetic infection. "Scientists say the Earth is more than a few thousand years old? Well, what would you expect them to say? This only proves they are evil and that all their so-called evidence is just lies." This is not a quote that, to my knowledge, has actually been published, but it is the clear meaning behind some actual books that have been published for use in American Christian home-school science instruction. Another meme in this set is the one that says you should be intolerant, even hateful, toward people who believe differently, especially apostates who have left your religious sect.

How could such ridiculous memes have spread? Dawkins (following on ideas from evolutionary biologists Helena Cronin and Amotz Zahavi) says that successful religion memes are those that prove the faithfulness of the believer. If all you believe is that God wants you to love other people, why, almost anybody can do that. But if God wants you to believe that you have to prove your faith by shooting a doctor who performs abortions, right out in public while he is attending church (as happened in Kansas in 2009), now, that is a little more difficult. Very few people have demonstrated this much devotion to their religion. Dawkins calls this the principle of costly authentication. There seems to be a self-reinforcing cycle of costly authentication of religious zeal among the Islamist suicide bombers.

Parasitic Religious Leaders

The third kind of response of a child to an extremely religious parent is for the child to follow in the footsteps of that parent. This is what evangelist Marjoe Gortner did, and continued to do, even after he knew that his parents' "ministry" was just a way of getting money and the admiration of followers. This is what author and film director Frank Schaeffer did, following the fame of his father, Francis, a theologian. He was particularly fond of writing conservative political screeds, as he described them. He later rejected this approach to religion.[13] Herbert W. Armstrong convinced the Worldwide Church of God that his theology, about the coming utopia of God on Earth, headed by the United States and the nations of Western Europe, was uniquely correct, and the church followed him. His son, Garner Ted Armstrong, used the same theology and had amazing charisma (on radio and television) besides. They got the church and its college to buy them a private jet, and they flew around the world to tell heads of state, everyone from President Suharto of Indonesia to the Duchess of Grand Fenwick, that God was going to usher in a utopia. Garner Ted used his influence for sexual exploits as well. In my early teens, I thought about becoming a follower of the Armstrong cult. One of my uncles (unknown to me until after his death) did so. Tulsa, Oklahoma, is famous not only for the late Oral Roberts, who openly admitted that his ministry was a very profitable business, but also for the way his son Richard notoriously abused ministry and university funds for private pleasures.[14] I am currently a resident of Tulsa. I can't seem to get away from these guys.

These are clear examples of people who parasitized the religious passion of their followers. In some cases, the parasitic leaders may be sincere in their delusion. In other cases, they may be in it solely for wealth and admiration, like Sinclair Lewis's fictitious Elmer Gantry. Some, like Oral Roberts, may be a little of both. Oral Roberts considered the wealth he amassed from his followers (including my wife's grandmother, who, like many others, sent him almost everything she could spare) to be evidence of God's approval. Jesus died poor, but many religious leaders appear to believe that the rules have changed.

Religious memetic parasitism began nearly as soon as religion did. The Cro-Magnon caves contained drawings and hand paintings in their deep recesses. But in the outer caves, the chambers not far from the entrance, artists painted their most extensive artwork. David Lewis-Williams interpreted these entryways as being where the tribal members assembled to listen to the stories of their religious leaders. It must have been quite an experience: paintings of huge animals, still stunning to observers today, revealed then concealed by flickering lamplight or torchlight, with ecstatic voices ringing and echoing off the cave walls, accompanied by wind instruments and percussion. Lewis-Williams said that it was not just the hills but the caves that were alive with the sound of music. This pattern of cave art can be found even in Chauvet, the oldest of the major Cro-Magnon caves (32,000 years old). There can be little doubt that tribal leaders used religion as a way of gaining power.[15]

Agriculture allowed civilization to begin. A slave could raise enough food to feed more than just his own family, so others were free to be full-time priests, artists, and warriors, all roles that were less professionalized in tribal society. The leaders of the earliest civilizations, including Egypt, Sumeria, and perhaps also Çatalhüyük on the Anatolian peninsula, proclaimed themselves to literally be gods. Later civilizations, such as the Israelites, considered their leaders to be chosen by God. The use of religion for power is very nearly the story of human history. When a genius named Thomas Jefferson proclaimed that the federal government could not establish a religion, it was truly one of the most radical departures ever made from the direction of history.

Religious leaders have parasitized religion not only for personal profit and pleasure but also for influencing political opinions of their followers. The most widespread examples are of televangelists telling their followers to support the conservative wing of the Republican Party, starting with the late Jerry Falwell and the Moral Majority in the 1980s. Pat Robertson has enough followers that he can maintain his own broadcasting network. He announces all of his own opinions as if they come straight from the mind of God, and none of his followers dares to dispute him. A typical example of one of his rants in 2008 (which I had to listen to in the hospital waiting room before a colonoscopy) can be summarized as "Send money send money send money bomb Iran send money."[16] As I was writing this chapter, a powerful earthquake brought incomprehensible misery to the nation of Haiti. The dead—those who could be pulled from the rubble—were piled along the same streets in which injured survivors writhed in pain. Some survivors covered themselves with rubble, only to be run over by vehicles whose drivers did not see them. Not even a day had passed when, on January 13, 2010, Pat Robertson announced that this earthquake was God's punishment on the Haitian people. They had, he claimed, made a pact with the devil two hundred years earlier that they would worship the devil in return for having the French colonialists ousted. I marvel that Robertson's thousands of followers believed (or at least made little objection to) even this most outrageous and hateful statement, which was clearly intended to undermine political support for the Obama administration's relief efforts.

A Recurring Pattern

There is a clear recurring pattern in the history of religion. Even when a religion is founded by a sincere person who has no aspirations to wealth or power, this person's followers quickly establish a religious hierarchy.

The example best known to most of us is the Christian Church. Jesus of Nazareth owned nothing; he had to depend on his followers. He had the most tenuous of hierarchies: twelve disciples and seventy assistant disciples. He also did not say very much (in the quotations attributed to him) about theology. After his death, "The Twelve" followers quickly established a governing council in Jerusalem, led by Simon Bar-Jonah (whom Jesus had nicknamed "Rocky" (Peter)). Their theological proclamations were fairly minimal, such as telling the church members to stay away from sexual sins and from blood. They did not like being pestered by an outsider named Saul of Tarsus, who renamed himself Paul and claimed to be an apostle like The Twelve. Paul went around the Mediterranean world preaching and found that there was a tremendous variety of religious practices in the different, independent Christian congregations. His letters, especially to the Corinthians, represent powerful attempts to bring ecclesiastical order into a chaos of religious fervor, but it was not until the fourth century that Christian doctrines were standardized into creeds.[17] Only after Roman emperor Constantine recognized the power of religious memes did he make Christianity into a church that had authority over people, and he often abused that authority.

Dissatisfied with abuses of power and the deadness of a religion that was used only to dominate the minds and spirits of people, Francis of Assisi began his own movement. Within a century, the religious order started by a man who rejected all possessions and wealth became as abusive and acquisitive as the ones preceding it.

The Reformation was a rebellion against abusive Catholic authority, but soon the Protestant churches also became tools of repressive governments. A group known as the Puritans left England to escape from the authority of the Anglican Church and found freedom of religion in the Netherlands. But freedom for everybody is not what they wanted. They wanted the freedom to impose their religion by law. So they sailed to Massachusetts, where they established a religious dictatorship that sometimes massacred the native inhabitants. Pilgrim leader William Bradford describes the way the colonists surrounded a Pequot village at sunrise. They set it ablaze and killed anyone who fled. Bradford wrote, "It was a fearful sight to see them thus frying in the fire and the streams of blood quenching the same, and horrible was the stink and scent thereof; but the victory seemed a sweet sacrifice, and they [the colonists] gave praise thereof to God, who had wrought so wonderfully."[18]

Religion is still the fountainhead of political power in many places. The Islamist (as opposed to benignly Islamic) movements, such as the Taliban and al-Qaeda, are examples too well known to require description here. Fundamentalist Christianity is famously powerful in American politics, as well. Many people assume that the Constitution created a wall of separation between church and state, but this is not quite the case. The First Amendment states that Congress cannot establish a religion, but a state, apparently, can. If a state did so, the federal government would immediately cut off its funding. But in many states, fundamentalist Christianity remains the unofficial religion. In my home state of Oklahoma, religion is the single-most important deciding factor in political elections, and people with extremely repressive religious beliefs are repeatedly elected to state office. At the state university where I work, the major official functions begin with overtly Christian prayers, and even our Muslim faculty members seem to accept this unalterable fact.

RELIGIOUS MEMES AND EVOLUTIONARY SUCCESS

Something as apparently dysfunctional as fundamentalist religion would seem to have nothing to contribute to evolutionary success. But clearly any tribe that had a stronger religion could prevail in war over a tribe with less religious zeal. A tribe of religious zealots could always whip a tribe of religious philosophers—the Stone Age version of Unitarians and Quakers. All the people in the prevailing tribe, and all the genes in their bodies, would likewise benefit.

And there is another possible reason that religious memes have proliferated. The human mind desires the experience of beauty. While religion memes have fed the worst of human feelings and actions, they have also fed the best of them. We have a natural, and probably irresistible, passion for the beauty of nature and for its green and flowering and chirping inhabitants. To scientists, this is an evolved human emotion Edward O. Wilson called *biophilia.*[19] To pagans, it is the power of the gods and goddesses of earth and forest and ocean. To monotheists, it is the presence of God within the observer. Religion is not the only medium for this feeling, but it is one of them.

It is easy to be inspired by the beauty of interesting landscapes, mountains, and waterfalls, by soaring birds and leaping whales and teeming jungles (so long as you are not actually walking through them and swatting the mosquitoes and pulling off leeches that fall on you like rain). But biophilia is an important force in helping people to love the places that are hard to live in. Consider these examples.

I taught a class at a field station one summer. One of my students was a young lady who grew up in western Kansas. The entire landscape of western Kansas is utterly flat and devoted entirely to wheat. Granted, there are beautiful places in Kansas, but this was not one of them. I had just driven through this young woman's hometown to arrive at the field station. She said, regarding western Kansas, that it was the most beautiful place in the world.

I grew up in the San Joaquin valley of California, which was beautiful at one time, before all its natural wetlands were drained and almost all its oak forests were cut down in the twentieth century, transforming it into farmland. In my childhood it seemed inexpressibly beautiful. It is a feeling I still get, if only a little, whenever I visit it.

Before we dismiss believing in the beauty of western Kansas or the San Joaquin valley as delusion, let us briefly consider what humans have had to endure. Much human misery has come from the seeming indifference or even hostility of the environment. Floods, volcanic eruptions, droughts. But the successful humans were those who loved their lands zealously, no matter what nature did to them, and endured. Their religious memes may have told them that there were evil spirits in the storm winds, but the Great Spirit prevails to make the world into a cosmos in which we can continue to live. We love the land in which we grow up. This is biophilia. Biophilia has been one of our adaptations to survival. Enshrined within the memes of some religions, biophilia has helped to keep us going.

RELIGION AND THE LIFE OF EARTH

The evolution of intelligence may have been inevitable. Evolutionary scientist Simon Conway Morris thinks so.[20] If a planet with earthlike conditions has a long enough period of stability (something vanishingly rare; see chapter 1), some form of intelligence will evolve. It has done so repeatedly on Earth, as Conway Morris points out. Dolphins evolved intelligence independently from humans, and it is a different form of intelligence, one we may never understand. The human lineage evolved intelligence at least three times. In Europe, the descendants of *Homo ergaster* evolved into big-brained Neandertals. In Asia, the descendants of *Homo ergaster* evolved into *Homo erectus*, which had brains just below the modern human range. In Africa, the descendants of *Homo ergaster* evolved into us. The evolution of altruism, as of intelligence, may be inevitable; our only evidence to the contrary is that science fiction can create sinister intelligences. Sexual selection may be inevitable. All the physical components of religion may be inevitable. Perhaps it was inevitable that religious memes brought all these things together into an amazingly powerful force.

The force of religion has governed the courses of nations and led to migrations that have opened up continents to civilization, as well as decimating the native inhabitants of those continents. It has literally inspired humans to do things that have transformed and degraded the planet—more so than perhaps any other force or adaptation. The underlying brain processes are here to stay, but the time has come to get rid of some of the memes. If we take charge of our minds, to the extent that we can, we might be able to get rid of the bad religion memes and keep the good ones.

Perhaps the most powerful meme is the idea that humans are the masters of the planet. This meme is most predominant in Western religions. True, countries dominated by Buddhists and Hindus also have environmental problems. The Ganges is one of the most polluted rivers in the world. But at least people in the Eastern religious traditions do not think that it is the will of God when they cut down forests or pollute the waters. The Hindus believed that the Ganges makes everything that enters it, even corpses, pure—and their religion made them ignore the obvious stench of cognitive dissonance. Frans De Waal speculates that human cultures that evolved in tropical regions could not invent a religion that placed humans, and humans alone, at the apex of creation, since they were surrounded by very humanlike apes and monkeys. Perhaps only in desert regions, where Jews and Arabs dwelt with obviously less intelligent cattle, goats, sheep, and camels, could a religion of human superiority have evolved.[21] A religion based solely on the words of Jesus of Nazareth would more closely resemble an Eastern religion than the hierarchical and oppressive Western religions supposedly based on them.[22]

Christianity is the Western religion that has created the most environmental destruction. For most of Christian history, and in most of Christendom today, the followers believe that the first chapter of Genesis has given them permission to subdue the Earth, to transform its forests and plains into cities with churches.[23] Perhaps most Christians choose to believe this, rather than the second chapter of Genesis, in which God tells Adam and Eve to take care of the garden in which they live. Most Christians have treated the Earth as if they were its father, disciplining it, rather than as if they were its mother, nurturing it. Christianity will remain a force of environmental destruction unless or until it embraces an idea that was best expressed, I believe, by Jerry Deffenbaugh, a Disciples Church minister. He said the Earth needs some mothering; it has had just about all the fathering it can take.

Chapter 8

SCIENCE

Humans are the mammals that never grow up.

Young mammals are endlessly playful. This trait allows them to try everything out and see how the world works; if they are lucky, they do not die in the process. Through trial and error, mostly error, they develop knowledge of the biological and social world and a set of skills to survive and reproduce in it. By the time most wild mammals have grown up, they have a set of behaviors that stay pretty much the same for the rest of their lives. Highly social mammals, whether apes or meerkats, continue to refine their knowledge throughout life, as social relationships change, but their adult learning is far more restricted than their juvenile education.

Humans are the major exception to this pattern. As explained in chapter 5, we are neotenous. In many respects, human adults resemble the juveniles, not the adults, of other ape species. Like young chimps, adult humans have big heads, big eyes, and are relatively hairless. And like young chimps we maintain the playful use of trial and error to keep learning about the world. Interestingly, the same can be said of our pets. Our dogs and cats never quite grow up, either. Old dogs do learn new tricks; they are not as playful as puppies, but they are much more playful than old wolves.

It is our big brains that have allowed us to make astonishingly productive use of neoteny. In previous chapters, we encountered evolutionary forces (sex and altruism) that led to our big brains. Another such evolutionary force, which we explore in this chapter, is the ability to figure things out. Like sex and altruism, figuring things out rewards us with pleasure. Once the human brain had reached its modern size by about a hundred thousand years ago, social and memetic evolution refined our ability to playfully explore, and redesign, our world.

In recent centuries, this ability has taken on a new form: science. Science, which is an old ability pressed into a new structure, has allowed us to understand ourselves, Earth, and the universe. Intelligence did not evolve for such a purpose, but it has ended up serving this purpose. It has been one of the most significant developments in the history of the life of Earth.

SERIOUS PLAY

The line between play and serious work is very thin. A playful kitten is learning to be an efficient predator. A juvenile chimpanzee playing with a stick discovers a new way to get termites out of a hole and eat them. And a human adult playing with a rock becomes the inventor of the first stone tool or improves the knapping technology of stone tools. The human may even use tree sap and fibers to attach a sharpened stone to a stick, play around with it, and end up inventing the spear. The process is quite simple. The intelligent animal thinks, "I wonder what happens when I do this," though not perhaps in words. If the experiment is successful, the animal remembers how to do it. If not, the animal remembers to not do it again. This is the rudimentary origin of science.

Inventiveness did not get off to a particularly rapid start in our species. Almost two million years ago, humans (of the species *Homo ergaster*) began to knap stone tools on both sides, thus inventing the Acheulean technology. As noted in chapter 5, this technology remained essentially unchanged for a million years. The pace of change did not greatly increase until Neandertals and modern humans independently developed advanced stone tools when the modern humans (Cro-Magnon) arrived in Europe from Africa. The Cro-Magnon Gravettian technology replaced the Aurignacian, the Solutrean replaced the Gravettian, and the Magdalenean replaced the Solutrean, each with improved tools, all in the space of about fifteen thousand years. The rapidity of this technological advancement was perhaps a Cro-Magnon response to their encounter with the very different and disturbingly intelligent Neandertals. Neandertal technology improved also, from the Mousterian technology that they had prior to the arrival of modern humans to the Châtelperronian technology that they had soon afterward—perhaps their response to the arrival of the Cro-Magnon.

There have always been two lines of thought among thinking apes: the conservative "If it ain't broke, don't fix it" line, and the progressive "We can do better" line. The conservative line may have prevailed in human populations that were not threatened by other human populations, but when *Homo sapiens* encountered *Homo neanderthalensis*, the time was ripe for innovation. And ever since that time, innovation has not slowed down. It has, in fact, reached a feverish pace, especially with the almost instantly obsolete digital technologies.

As explained in the preceding chapter, humans inevitably want to understand the sensations that come into our big brains. All intelligent animals do this, but humans make use of a lot of information, as well as abstract concepts created by no other species. We want, therefore, to understand how all these things fit together into a coherent system. Instead of many disconnected little experiments, in the manner of a kitten, we carry out complex sets of experiments, in which the result of one experiment leads to the next one, forming a network of understanding.

The Social Network of Knowledge

The network of understanding soon became so large that nobody could remember all of it. A social network, however, can store more information than any individual. When language evolved in the human species, humans were able to share information and concepts with one another and collectively preserve a large body of knowledge. Much of this knowledge was necessary for survival. Hunters and gatherers have to know a huge amount about plants, animals, weather, and ecology—certainly much more than most civilized people know about nature today. Therefore when a person begins a new experiment, he or she starts from a knowledge base much larger than his or her own. With the advent of village life in places such as Çatalhüyük (in modern Turkey) and Jericho (in modern Israel), and then civilization on the Akkadian plain, there was even more knowledge, and it was not long before writing was invented to store it.

The Primitive Origin of Science and Technology

None of this is particularly surprising. We call it common sense, or the process of elimination. We try something; if it works, we keep it, and if it doesn't, we discard it. Centuries of this process led to complex technologies such as agriculture and to the first sciences such as astronomy. Even before village life had become civilization, people built technological marvels such as Stonehenge, which apparently had no utilitarian purpose. The rocks of Stonehenge were instruments with which to study the patterns by which the sun and stars moved (and then worship them).

All this was based on an unspoken assumption that there were laws of nature and that these laws did not change. If the tribesmen of what is now the Salisbury Plain of England had believed that the sun moved in an unpredictable fashion, they would never have built Stonehenge. Greek mythology said that the sun was Apollo driving his chariot across the dome of the sky, and that the gods were often moody and unpredictable. But the ancient Greeks never doubted that Apollo's chariot always, always reached its lowest arc on the day we call the twenty-first of December, regardless of what mood he might have been in. No matter what Trickster Coyote might do in Native American mythology, spring always came, following the same sequence of events. Warm and cold days were often unpredictable in North America, but when the silver maple and elm buds began to open, spring was near. Belief in "laws of nature" allowed the primitive origin of science.

Organized Common Sense

Science is a disciplined version of the process of elimination. Thomas Henry Huxley, the scientist who defended Charles Darwin's ideas in public, called science "organized common sense."

Nearly everyone uses a scientific way of thinking on an almost daily basis. For example, if the oil light comes on in your car, your brain switches to a scientific mode of thought. Your first idea might be that your car does not have enough oil. Then you test this idea. You check the oil level with the dipstick. If the oil level is low, you get more oil. But suppose the oil level is not low. Your next idea may be that the oil pump is defective. So you have a mechanic check your oil pump. If that is the problem, you just get a new oil pump. But suppose the oil pump is working. Perhaps your next idea is that the oil filter is clogged. So you have your mechanic check the filter. And if that isn't the problem, then perhaps the oil sensor itself is malfunctioning. What I have just described is the scientific method. You have an observation that you need to explain. You propose a hypothesis, and then you test it. You may end up testing a series of hypotheses. Eventually, unless the problem becomes so complex that you decide it is not worth the time and money to solve it, you will reach an answer. It is common sense. The process of elimination. Science. Commonsense memes have been around a long time. They have recently been formalized into a meme-set known as science.

Common sense also allows you to avoid various pitfalls of interpretation. For example, you would not conclude that all oil sensors are defective, just because yours is. You would need to test lots of oil sensors before you drew a general conclusion about them. A scientist would say that you need an "adequate sample size." Common sense also tells you that you need to obtain your sample of oil sensors from lots of different kinds of cars, not just one, in order to draw a conclusion about oil sensors in general. If Edsels often had defective oil sensors, you would not conclude that the oil sensors in Toyotas are also defective. A scientist would say that you must make sure your observations have "external validity." Common sense also tells you that you need to make sure your observations are actually telling you what you want to know. For example, if you check the dipstick while the engine is running, it will not give you a valid measurement of oil level. A scientist would say that you must make sure your observations have "construct validity." So, what we scientists are doing is using a commonsense approach to answering questions and solving problems, and using fancy terms to describe it so that you think we are smart.

That is what Huxley meant by organized common sense. The scientific method is a series of constraints that prevent your thinking from becoming lazy and straying away from common sense. Science is a yoke. A yoke constrains the oxen, but it allows them to move the cart of knowledge forward through the mud of human confusion. Using common sense at its best, scientists are always trying to think of things that may have gone wrong, mistakes in reasoning that may have led them to misinterpret their results. Any one scientist might overlook a flaw in reasoning, but set a bunch of scientists on the problem and one of them will nearly always detect an error that the others have made. It may take decades, but the error will emerge and get corrected. Commonsense trial and error led to the invention of agriculture, aqueducts, and cathedrals. Science has made common sense into a profession. Everyone can do this kind of thinking, but scientists are trained to do it very well.

By understanding how things work at the present time, scientists can make reasonable predictions about the future. Once again, this is something that anybody can do, but scientists are good at making predictions based on lots of information. Scientists (and economists, who are also scientists in a general sense) know that you cannot predict the future of the economy based on just the last week of stock market data.

But there are some rules to which scientists strictly adhere and which many other people sometimes fail to follow. The first is to not trust hunches. A hunch is an untested hypothesis, which is no good until it passes the test of evidence. Intuition, "following your gut," sometimes works because your brain is subconsciously analyzing the data. A hunch is often a good place to begin, but it is not a good place to end.

Another rule is not to make assumptions. Sometimes assumptions are unavoidable. But scientists specialize in questioning the assumptions behind our understanding of the way the world works. Good common sense and good science demand that we not believe things just because they have always been believed.

Another rule is repeatability. If something happens only once, you cannot investigate it either by common sense or by science. You may, however, be able to study evidence associated with it, as I shall explain later in this chapter.

Scientists do not always just follow rules. They have to be ready for surprises. Arno Penzias and Robert Wilson did not start out to find background radiation from the big bang (see introduction), but when they found cosmic radiation that they could not explain, they used the scientific method to figure out what it was.

A final rule of both common sense and science is that your observation have a physical cause. It would not have occurred to you that the oil light came on in your car because a demon made it come on. What scientists do is to rigorously apply this rule to the whole realm of nature and of human experience. Many religious people assume that the complexity of nature had to be designed by a supernatural force. Scientists do not accept this, and instead seek natural explanations such as evolution. Many religious people assume that when a person says or does something bad, it is because the person's spirit is evil. Scientists look for possible brain malfunctions or negative childhood experiences. Science is common sense applied to everything. Many religious people accept science up to a point, but they make sure that science does not intrude on their religious beliefs. Most scientists are not, however, afraid to enter the territory of testing religious beliefs. And when they do, it is sometimes fun to see what they discover—sort of like four teenagers and a big dog in a van discovering that a ghost is really a man trying to scare people.

And it is this last rule that has caused the most conflict between professional scientists and other people.

TENSION BETWEEN SCIENCE AND RELIGION

Evolution produced human minds and a collective human mind, with capacities both for religion and for a primitive form of science. As civilizations rose and fell, these two adaptations remained in continual conflict. The reason is that the leaders wanted to use religion to keep the minds of their subjects under control, while the process of reasoning caused people to think for themselves. The people who used reasoning the most—the philosophers and scientists—were often very religious people, but their religion was not the kind that told them to believe whatever the kings and priests told them to believe. Their religion was more like that practiced by Jesus of Nazareth. Instead of just repeating what the religious leaders said, Jesus went out into the countryside and observed for himself the lilies of the field and the birds of the air. He expected to find religious insights for himself and to find them outside the temple.

It is therefore not surprising that the beginnings of science and of freedom of thought occurred together. Mathematics and philosophy grew in the relative freedom of Athenian democracy. Philosophers and scientists managed to have a certain degree of freedom of thought even within repressive societies. Antoine Lavoisier practically invented modern chemistry within the repressive French monarchy, but the Sun King did not tell him what to believe about chemistry. Scientists in societies with repressive governments could often think freely, so long as they did not talk about it. The heliocentric theory of the Polish scholar Nicolaus Copernicus was not published until shortly before his death. When scientists did talk about their ideas, conflicts emerged, such as the trial and house arrest of Galileo Galilei for his scientific conclusions that Earth revolves around the sun and that moons revolve around other planets, such as Jupiter.

And it is little surprise that scientific research thrives most in societies with the greatest freedom of thought. For the most part, the United States and Western Europe have been open to free scientific inquiry. The strident American battles over evolution do not so much concern scientific research as they do education. In stark contrast, biological research was forced into erroneous directions in the Soviet Union under the dictator Joseph Stalin. Trofim Lysenko, a plant breeder, claimed that one species of plant or animal could be transformed into another if you forced it to develop under different conditions. All available biological data contradicted this. But Lysenko's ideas resonated with communist ideology: if you forced people to live in a communist regime, they will change into good communist people, so why should this not also be true of plants and animals? Scientists who disagreed with Lysenko—such as the geneticist Nikolai Vavilov, who was respected by his peers around the world—were imprisoned, and some, like Vavilov himself, died in prison. There have not been very many scientific martyrs, but Vavilov was one. Genetics research in the Soviet Union suffered decades of worthlessness as a result.[1]

Does Religion Stifle Science?

Scientific research thrives among people whose religion permits free inquiry. One comparison will serve as an example: the contrast between conservative Christians and liberal Jews.

There are many scientists who are conservative Christians, but not very many outstanding ones. It is as if conservative Christianity acts as a cage that restrains scientific thought—a cage, not a yoke. This need not be the case. Everyone cites Francis Collins, former director of the Human Genome Project and current director of the National Institutes of Health, as an example of a brilliant scientist who is a conservative Christian.[2] But there are not too many others like him within conservative Christianity. And Collins is no fundamentalist. He will openly tell you that the DNA he studies reveals an evolutionary history for all organisms.

As far as I am aware, there are no Christian fundamentalist scientists who have done outstanding scientific research. There are, in my opinion, two reasons for this.

One is that fundamentalist scientists often restrict themselves to areas of science that do not conflict with any church doctrines. Back when I was a fundamentalist, I earned a PhD in botany, but my research (concerning plant responses to light and water) was on a topic that no church had any proscriptions against. Scholars at conservative Christian colleges, such as the one I worked at right after completing my doctorate, are very careful to not think outside of the constraints that have been placed on them by their church leaders. This should not, however, be too difficult to do. There are many areas of science that would appear to be neutral with regard to religion. Examples include any of the medical sciences, such as human physiology or epidemiology. While it is true that ancient scriptural sources attribute human diseases to God or to demons,[3] few even among the fundamentalists still believe this. There are no campaigns, as far as I am aware, to have demonology incorporated into medical school curricula. Therefore, fundamentalist scientists ought to pursue research in these areas just as well, and as vigorously, as scientists who are not religious. Oddly enough, fundamentalist scientists are less productive of these kinds of research as well. I do not know the reason for this, but it is possible that conservative Christian scientists are complacent because they think that all the important questions in life have already been answered by their religion. Where are the creationist bacteriologists? The creationist plant physiologists? There must be some, but they are obscure.[4]

The other reason is that fundamentalist scientists have a zeal for creationism and wish to use their scientific training to defend it. Because creationism is wrong, these efforts fail. Sometimes in their zeal to twist the facts of science around their creationist framework, they invent new religious stories that are not even in the Bible.[5]

I did not last very long in conservative Christian higher education. I continued to work as an adjunct professor with a Christian college that was willing to embrace modern science and environmentalism. But aside from that, I was glad to leave the world of conservative Christian higher education.

In contrast, most Jews have great respect for their religious tradition and believe in some form of God, but very few are fundamentalist. Judaism does not restrain scientific thought the way conservative Christianity does. Jewish theologians have a long tradition of expanding biblical principles in new directions. In effect, a devout Jew often argues with God, just the way Abraham and the prophets did. Also, there are many secular Jews, while "secular Christian" is considered an oxymoron. It is therefore no surprise that the list of Jewish Nobelists is very impressive, while the list of conservative Christian Nobelists is short. Wisdom is highly prized in Jewish culture, while a devout Christian is often encouraged to drain out his own thoughts and simply let the thoughts of God (or the thoughts of preachers who appoint themselves as God's surrogates) pour in.

A Revolution of Science and Freedom

The leaders of the American Revolution absolutely believed that freedom of thought required both political and religious freedom. The American Constitution established the first, and the Bill of Rights established the second.

One of the greatest scientists of the eighteenth century was a man who allowed his genius to make full use of both political and religious freedom. Benjamin Franklin was not only an entrepreneur and inventor (whether of practical things, such as improved methods of printing or of artistic things, such as an improved glass harmonica), but he also made breakthroughs in scientific understanding. He was always observing and thinking about what he observed. As noted in chapter 1, Franklin noticed that 1783 was a cold year in Europe; he heard that the volcano Laki had erupted in Iceland; he knew that particles, such as volcanic dust, block sunlight. He drew the conclusion that volcanic eruptions can cause cool weather—and may have been one of the first people to do so.

Franklin also noticed that static electricity and lightning had a similar appearance. Rather than dismissing this as mere coincidence, he tested his idea experimentally by means of the famous kite experiment in which lightning had the same effect as electricity in causing a key to glow. He went further and created an explanation for how electricity worked: he invented the concept of electric current. Finally, he took a step that most scientists are not lucky enough to take: he used his science to invent something very important—the lightning rod. By diverting electricity from lightning into the ground, his invention has saved countless thousands of buildings from burning down.

Thomas Jefferson had a strong interest in science and invention. In 1791, Congress (recognizing his scientific interests) requested that Jefferson inquire into Jacob Isaacks's patent claim for a new desalination process. Jefferson also discovered some mammoth bones and formulated the hypothesis that mammoths, though extinct in Virginia, were still alive in the American West, perhaps in the Louisiana Territory that the United States had purchased during his presidency. Scientific inquiry (including a search for mammoths) was a major motivation of the Lewis and Clark expedition. Alas, the explorers found no mammoths. Meriwether Lewis sent many plant specimens back to Jefferson, who apparently examined them. With founding fathers like Franklin and Jefferson, who recognized that investigators must be free from intellectual restraint, the United States became a world leader in scientific research and invention.

The United States is, and always has been, a very religious country. However, intellectual freedom has flourished despite religion, not because of it. It has flourished because it is not legal for religious groups to prevent scientific inquiry. Religious groups can, however, brainwash their children to automatically reject anything that does not conform to their church doctrines. I see this in the classes I teach at a university in rural Oklahoma. Many of my students come to college already convinced that science professors are atheists and that they should not believe what they hear in science classes. Local school boards in rural areas often pressure teachers to skip evolution, even when it is a topic required by state educational authorities. America continues to produce many good scientists, but they are usually individuals who have chosen to reject the constraints of their religious background, or who come from families or from parts of the country where these restraints are less powerful.

Science and religion are both means of doing great harm or doing great good. There is an important difference between them, however. Fundamentalist religion assures its zealous adherents that they possess inerrant knowledge—that is, they cannot be wrong, and all who disagree with them are damned. Science does not do this. Scientists know that a new set of facts can upset a previously held theory. This is unlikely to occur, of course. For example, no single fact could make evolutionary science collapse. Evolutionary science was built up from many hypotheses, each confirmed by facts, and it would take many contradictory facts to tear it down. But any scientist would admit that this could conceivably happen, however unlikely it might be. As a result, scientists in general are more humble than religious nonscientists when it comes to beliefs.

Scientific Study of Religious Claims

Religious claims have seldom been subjected to scientific investigation. One such investigation was conducted thousands of years ago by the Israelite leader Gideon. According to a story in the sixth chapter of the Old Testament book of Judges, Gideon believed that God had told him to take a small band of Israelite warriors against a large army of enemy soldiers. But Gideon wanted to make sure, so he laid a fleece out on a threshing floor. Gideon then told God to confirm his words by making the fleece wet while the floor was dry. The next morning, behold, the fleece was wet and the floor was dry. Gideon then wanted to replicate his experiment. He put out another fleece, and told God to confirm his words by making the fleece dry while the floor was wet. The next morning, lo, the fleece was dry and the floor was wet. All the while, Gideon kept apologizing to God and begging God to not get mad at him for attempting an experimental confirmation of God's activity. While Gideon's line of reasoning had faulty construct validity ("The fleece was wet, therefore God really did tell me to lead an army into battle"), it was one of the few religious pronouncements, ancient or modern, that was not based solely on the claims of a religious leader.

More recently, scientists and theologians have attempted to investigate whether religious activities such as prayer actually have an effect on the physical and human world. Consider, for example, intercessory prayer, in which some people pray that God would heal a person who is ill. Sometimes, when people pray, the person for whom they pray recovers from the illness. What might cause this? One possibility is coincidence. Perhaps the person would have gotten well anyway. Another possibility is psychosomatic healing. *Psychosomatic* means that the brain (*psyche*) affects the body (*soma*). If you give a patient a pill, the patient will often begin to feel better because his brain tells him that he ought to feel better. This could lead to the false conclusion that the drug works when it actually does not. To control for this, researchers nearly always test drugs by giving some of the experimental subjects the real drug and others a placebo (sugar pill) that looks like the real drug, without telling them which pill they have received. If the drug is effective, the patients who receive the drug will recover more quickly or more often than patients who receive the sugar pill. The sugar pill almost always works better than nothing at all.

Sometimes the placebo effect backfires. Patients in experimental drug studies know that drugs usually taste bad. Therefore, if they receive a pill that does not taste bad, they assume it must be a sugar pill and they may subconsciously make themselves feel worse as a result. The good-tasting pill disappoints the patients. These results might lead to a false conclusion that the drug works when it actually does not.

Taking the placebo effect into account, prayer could act like a drug: the patients who receive intercessory prayer, *and who know they are receiving it*, might start to feel better simply in the knowledge that people care about them, regardless of whether the prayer itself is having any effect.

Perhaps the most famous scientific study of religious claims was published in the *American Heart Journal* in 2006.[6] The question addressed by the article was simple and profound: does intercessory prayer actually work? If the question had been stated as "Does God answer prayer?" it could not be addressed by science, which has no way of studying God. But scientific research *can* study the question of whether prayer has effects on the observable physical world. In this study, the investigators wanted to determine whether intercessory prayer would help patients recover from heart surgery. The initial idea was simple: There should be a higher rate of recovery among patients for whom people prayed than among patients for whom people did not pray. While earlier studies of the "efficacy" of prayer had not, apparently, taken the psychosomatic effect into account, the 2006 study was designed to specifically cancel out the psychosomatic effects of prayer. The lead author, Herbert Benson, is a medical doctor who works for the Mind/Body Medical Institute in Chestnut Hill, Massachusetts; psychosomatic effects are his specialty.

Two problems will be immediately obvious to any reader. First, it is not possible to actually prevent people from praying for the patient. You cannot tell the person's family or church to not pray for the person who is ill. In such a study, you cannot be sure that the patient "doesn't have a prayer." But what the investigators did was to arrange for *supplementary* intercessory prayer to be *added to* whatever background level of prayers might have been taking place.

A second problem was deciding which patients to include in the study. The team chose, as their pool of candidates, patients who had already been scheduled to receive a coronary artery bypass graft. Permission was first obtained from their doctors and then from the patients themselves. An important part of this investigation was to make sure there were enough patients and that they were representative of the American population (that is, the study had what I have previously called adequate sample size and external validity). This was in fact the case: the six hospitals were in Oklahoma City, Boston, Washington, DC, Memphis, Rochester (Minnesota), and Tampa. Initially, 3,295 patients were identified as eligible; 1,493 refused to participate, and the other 1,802 joined the study.

Were the investigators qualified to study such a broad question? Yes, because the panel of investigators included people with a wide variety of professions: the lead author (Herbert Benson) is a physician, and other members of the team included other physicians, nurses, public health experts, and even theologians. The study was partly supported by the John Templeton Foundation, which funds scientific studies of religious, particularly Christian, questions.

Who was going to do the praying? The investigators searched for religious groups from as wide a variety of backgrounds as they could. What they ended up with was just three groups: two Catholic monasteries and Silent Unity, a prayer group in Missouri that is based on Christianity but recognizes the validity of many different religious traditions. The researchers could not find any other groups that would agree to pray on a daily basis for several years for a list of people whom they did not know. Each patient was prayed for by an average of thirty-three intercessors every day.

In order to detect the presence of a psychosomatic effect, the investigators did two things. First, they divided the patients into three groups, not two. In groups 1 and 2, the patients did not know whether they were receiving supplementary prayer. Group 1, without the patients' knowledge, received supplementary prayer; Group 2 did not. Group 3 consisted of patients who received supplementary prayer and knew it (table 8-1).

Table 8-1.
Study of the Therapeutic Effects of Prayer

3,295 eligible patients
1,493 patients elected to not participate
1,802 patients randomly assigned to three groups

Group	Number of patients	Complication rate
Uncertain of receiving prayer:		
Group 1: Received prayer	604	52.2%
Group 2: Did not receive prayer	597	50.9%
Certain of receiving prayer:		
Group 3: Received prayer	601	58.6%

Table 8-1. The setup of the STEP study (study of the therapeutic effects of prayer) as described in the text.

Second, the investigators needed to make sure that their caretakers did not treat the three groups differently. If the nurses knew that a patient was in Group 3, they might take extra-good care of the patient, increasing the chances that the patient would recover. In such a case, the recovery would not be due to prayer but due to the nurses. All the investigators could do was to tell the patients to keep their status secret and not let their caretakers know which group they were in. As you can see, the investigators did everything they could to prevent factors other than the supplementary prayer itself, and the awareness of receiving it, from influencing the experiment.

As it turned out, the makeup of all three groups of patients was very similar: about sixty-four years old, about 70 percent male, about 90 percent Caucasian. About 15 percent were current smokers, and about 53 percent had been smokers. Their cardiovascular case histories were all very similar. In all three groups, about 80 percent of patients participated in religious groups, and the groups were all similar in their religious affiliations (mostly 60 percent Protestant, 30 percent Catholic). If any of the groups had been significantly older, more male, less Caucasian, more religious, or if any of the groups had contained more smokers, the research might have been invalid. Luckily, this was not the case. Perhaps most important, patients in all three groups were very similar in what they believed about the efficacy of prayer: About 65 percent of each group signed a statement that they strongly believed in spiritual healing.

Finally, what did the investigators measure? That is, did the study have construct validity? They measured any heart complication that arose within thirty days of the surgery, whether or not the patients remained in the hospital. They lost contact with a handful of patients, ending up with 604, 597, and 601 patients in Groups 1, 2, and 3, respectively.

And now, the results. The most important comparison was between Groups 1 and 2, the groups of patients who did not know whether they would receive prayer. In Group 1, which received supplementary prayer, there was a 52.5 percent postoperative complication rate. In Group 2, which did not receive supplementary prayer, the complication rate was 50.9 percent. These numbers might appear to show that supplementary prayer makes recovery about 1.6 percent less likely. But the two numbers are similar enough that the difference between them could have been due to chance. The conclusion was that supplementary prayer has no effect on the outcome of heart surgery recovery.

Then the results get really interesting. Postoperative complication rate in Group 3 (the patients who received supplementary prayer and knew it) was 58.6 percent, a significantly *higher* rate of relapse than in either Group 1 or Group 2. This result astonished the researchers. They found a psychosomatic effect opposite of what they anticipated: people who knew they were receiving supplementary prayer fared worse than those who did not know. The researchers had no explanation for this. Some people have interpreted this as performance anxiety: the people in Group 3 may have felt that, if they did not get better, then they would be tarnishing God's reputation as a healer. They were motivated to get better, to prove their religious views correct, but this may have made them more nervous and more likely to get worse.

These results left religious believers scrambling. Some said that you cannot make God participate in an experiment. You cannot manipulate God or turn God's power on and off like a spigot just by having people pray. Of course, the scientific study was not about God but about the efficacy of the prayer itself. It did leave the religious people in a bit of a bind, however: If they are correct, then apparently God refused to answer prayers because God was miffed that scientists were watching. God was mad at the heart study participants just as Gideon feared God would be mad at him. This seems petty and inconsistent with the image that most people have of God. As I noted in the previous chapter, the strongest religious memes are those that claim that *contrary* evidence actually constitutes proof of their correctness. Well, *of course*, God did not answer these prayers; he wants to reward only blind faith, not faith based on evidence. This belief, incidentally, contradicts an emphasis on evidence (such as the evidence that early Christians offered to support the resurrection of Jesus) that was central to the messages of Old Testament Jews and New Testament Christians.

Some believers claimed that the supplementary prayer did not work because most, perhaps all, of the patients were being prayed for. In other words, the supplementary prayer exceeded the amount that was necessary. Another example would be that plants grow more when you give them more light, until you give them all the light they can use; beyond that point, further illumination has no effect. However, about half of the patients died. Does this mean that intercessory prayer cannot be more than 50 percent effective?

Some fundamentalists claimed that the intercessory prayer was worthless. They claimed that God would not listen to the prayers of Catholic monks and nuns or to the prayers of an interfaith prayer group. God only listens to good Holy Roller fire and brimstone groups—none of whom, despite invitations, elected to participate in the study.

Meanwhile, the only conclusion the researchers reached was that prayer does not reliably work. And they admitted limitations to the study. Maybe intercessory prayer works, but just not on heart disease patients. Or maybe intercessory prayer works, but not enough to affect these results (that is, the efficacy is less than about 10 percent). Or maybe the people who did not receive extra prayer started dying, but it was more than thirty days later (after the end of the study). They admitted that all kinds of excuses could be made. And they concluded as follows: "Our study focused only on intercessory prayer as provided in this trial and was never intended to and cannot address a large number of religious questions, such as whether God exists, whether God answers intercessory prayers, or whether prayers from one religious group work in the same way as prayers from other groups."

This study also highlights a very important difference between science and religion. This scientific study reached a definite conclusion; the hypothesis (that prayer helps heart patients recover from surgery) was not confirmed. Religion, on the other hand, makes statements such as the one by famous Christian writer Clive Staples Lewis. Lewis said God always answers prayer, but the answer can be yes, no, or wait.[7] No conceivable set of observations or experiences could ever confirm or contradict such a statement. No matter what does or does not happen, a believer can claim it as evidence that the prayer was answered.

Progress of Science and Stagnation of Religion

Both religion and a rudimentary form of science are at least as old as *Homo sapiens*. The human species will never be without both of them. Science has rapidly developed into a complex and stunningly successful way of understanding the universe and manipulating the conditions of human life. More scientific breakthroughs have occurred in the last fifty years than in all previous human history.

Meanwhile, religion has more or less stagnated. When religion was the only game in town, the universe was small and revolved around Earth, and God miraculously created all things bright and beautiful, all creatures great and small, so that we could praise God. God also created dangerous and ugly creatures to keep the fear of hell in us. We believed that all our thoughts and feelings came from spirits that enlivened the lumps of clay that were our bodies. Science has dispelled all these beliefs. Theologians have attempted to bring religious thought more in line with what we now know about the world. In many cases, the result has been theological systems that make very diffuse and mystical claims. While science has explained more and more, religion has explained less and less about our entire experience of the world.

OTHER NONSCIENTIFIC MODES OF THOUGHT

Religion has been the most extensive, expensive, and long-lasting antiscientific process in human history. But it has most certainly not been the only one. Another example of an antiscientific idea is the belief in natural medicines. Just as, in earlier centuries, religious people rejected medical science (some, like the Church of Christ Scientist, still do), so do many adherents of the natural health craze. They think that one can remain healthy by buying and using very expensive powders and extracts from special plants and animals that they believe to have healing properties. It is almost a religion: they seem to think that some universal godlike force has infused the secrets of health and potency into the natural world, and our job is to go find them. This view contradicts evolutionary science as much as creationism does. Evolutionary science says that the chemicals found in plants and animals got there because they enhanced the fitness of the plants and animals that possess them, not because they might help us. Sometimes it is religious people who are also followers of natural medicines. I visited an Amish store where the owner told me all about the wonders of herbal medications, in particular of horehound, a weed in the mint family.

The natural medicines craze has its foundations in fact. Nearly half our pharmaceuticals were discovered in plants, and we have not found all of them yet. I myself am involved in research into the medicinal qualities of a species of plant. But such discoveries occur by chance. There is no law of nature that says that the cures for all our diseases are hidden in the botanical lore of rainforest tribes.

Centuries ago, scholars believed that plants could cure diseases of the parts of the body they resembled. For example, liverwort plants look like little green livers; therefore, liverwort powder could cure liver diseases. This belief has been totally abandoned by medical science but lives on in health-food stores. A particularly humorous example is the expensive Tribulus pill that supposedly enhances sexual potency. People who buy these pills may not know that Tribulus is the puncture vine, famous for seeds that have sharp spikes on them. Someone in the past saw these spikes and was reminded of penes, and assumed that eating the seeds would cure impotence. (And it sometimes works. If a person pays several dollars for a bottle of placebo pills made from Tribulus powder and then relaxes in the knowledge that he will no longer be impotent, perhaps his performance anxiety will go away and take his impotence with it.) American devotees of herbal medicine scoff at Asians who think that consuming powdered rhinoceros horn will enhance sexual potency, even as they gobble down their Tribulus pills. Herbal medicine, like religion, earns its professional practitioners a lot of money—not as much as the big preachers get, but more than most scientists earn.

Other examples of the denial of scientific inquiry include the widespread beliefs that vaccines and genetically modified foods are insidiously dangerous. The fact that rigorous scientific research has confirmed the safety of these two technologies only "proves" to opponents that there must be a conspiracy.[8]

It is easy for a scientist such as myself to denounce such beliefs. How could the human brain tolerate such things? But remember what the human brain evolved "for." It did not evolve as a way of understanding the universe. It evolved as a sexually selected entertainment system and as a way of navigating the complexities of social relationships. It evolved as a way of understanding just enough about the world to survive in it. The opponents of science are not stupid; they are just human. Considering that the human brain did not evolve for science, it is amazing to me that antiscientific thought is not more common than it is. It is more luck than we have any right to expect that the human brain was fertile ground for the memes of science.

SCIENCE AND THE FUTURE OF THE EARTH

The zeal of scientists is not less than that of religious people. Scientists bravely venture into jungles in search of new species and onto mountains to measure melting glaciers. To be a good scientist requires almost as much sacrifice of the self as to be a religious convert. As one cartoon depicted it, if a normal person presses a button and receives an electric shock, he concludes, "I guess I'd better not do that." If a scientist presses the button and receives a shock, he asks, "I wonder if that happens every time" and repeats the experiment. A scientist's thirst for knowledge can be so great that it can put him or her in danger. Even those of us scientists who live safely have sacrificed the possibility of lucrative careers in order to feed our hunger for understanding.

Some scientists have zeal only for understanding the universe. An example of such a person was the chemist Antoine Lavoisier, whose passion was chemistry. He was also an official of the French monarchical government and was responsible for oppressive taxation in Paris. When the Revolution came, Lavoisier went to the guillotine because of this taxation, not because of his science. When the judge said, "The Revolution has no need of scholars," he meant that Lavoisier's science was not enough to compensate for his oppression. Lavoisier's passion for science made him ignore the suffering of poor Parisians, and that was his undoing.

But many scientists have a zeal for using science to help protect the natural world and to help the people whose lives depend on it. Perhaps the best example is George Washington Carver. All his research was focused on how to improve agricultural productivity of poor farmers in the rural American South and how to increase the value of their products (such as peanuts and sweet potatoes). He was also a Christian, though not a fundamentalist.[9]

Science has transformed human life and therefore the life of the Earth, allowing people to understand, for the first time, how the universe actually works. In effect, it has allowed Gaia to understand "herself" for the first time. It is as if Gaia had been sleepwalking for billions of years, then awoke in a spasm—and within a few centuries, a heartbeat of time, she suddenly understood the universe, her own beauty, and her own peril.

Science unleashed the power of humans to damage the Earth, but, perhaps not too late, it has revealed the knowledge of how humans can live upon the Earth without damaging it. Paleolithic instincts dominate our human minds. Such instincts aided the success of Stone Age people, but so great is our population and technology today that those instincts will make us destroy the Earth if we obey them. The most successful liberation from our instincts has been science. Science by itself cannot save the life of Earth from disaster, but disaster cannot be avoided without it.

Chapter 9

FAITH IN PHOTOSYNTHESIS

The process of photosynthesis, which evolved about three and a half billion years ago, has kept Earth alive and provided the energy that allowed the other processes we have considered—such as symbiosis, altruism, and sex—to make Earth into the beautiful planet it is today. In this final chapter, we will look into the future, westward from the pedestal of rocks in Kansas. Earth faces many threats now and great uncertainties in the future, but we can have faith that nothing will stop photosynthesis from keeping Earth alive until the sun expands and the oceans boil away.

It is the fate of all organisms and all planets to die. But a lot of things will happen before the inevitable decline of Earth. Looking westward from Kansas, we see dusty skies and a landscape that we can only vaguely describe. Our human brains cannot resist wondering what this landscape will be like. For millennia, our brains have constructed scenes of eternal heavens and underworlds, models of our future. Today, we look to science to satisfy this same curiosity.

THE NEXT FEW THOUSAND YEARS

As Yogi Berra said, it's tough to make predictions, especially about the future. The ultimate future of Earth is inevitable, but there are many uncertainties about what will happen before that fiery end.

From the very beginning, the biography of Earth has been a story of good luck, as explained in the first chapter of this book. And these particular items of good luck will not change for billions of years. The Earth will remain propped up by the moon within a habitable zone of the solar system, swept clear of most asteroids and comets by Jupiter. This much we can predict with certainty. Earth has also had bad luck, such as supervolcanoes and asteroids. Such catastrophes seem unlikely to occur for at least the next several million years. Most astronomers believe that the asteroid that is headed toward Earth in 2029 (with a return visit in 2036) will miss our planet by a wide margin. In terms of chance events and the fate of Gaia, it looks like clear skies ahead. But many uncertainties remain regarding the effects of individual and collective human decisions on the health of the planet. Not many years ago, most people wondered whether nuclear holocaust might forever change the future of Earth. Nuclear war remains a threat, but not on a large enough scale to destroy or nearly destroy humankind. Other threats, no less significant, remain.

Population

One threat is the human population explosion. The world population has doubled during my lifetime, and I am not old. I gave a junior high graduation speech in 1971 about the overcrowded spaceship Earth—with three and a half billion people in the world! I could hardly have imagined that, in middle age, I would see the world population double that amount. And most of these people, even the poorest, are using more resources and creating more wastes than did the people of the past. Back in caveman days, it didn't matter to the Earth what anyone did; there were only about a million people. And when a caveman threw away a stone tool, it was just a rock; now we have billions of people throwing away batteries, which are toxic waste. Even the most altruistic people (I imagine myself to be one) generate lots of waste and have a huge impact (or "footprint") on the ecosystems of the Earth, even when we try to not do so. We altruists keep our thermostats down in the winter, use ceiling fans rather than air conditioners in the summer, drive little cars, and reuse scraps of paper, but we are still helping to kill the Earth.

The book *Famine 1975! America's Decision: Who Will Survive?* by William and Paul Paddock had many valid points but was spectacularly wrong about the date.[1] The reason famine did not grip the world in 1975 is that the intensity of agricultural production has increased more than the population. As a result, the proportion of starving people is actually lower on a seven-billion-person Earth than it was on a three-billion-person one. Much of the credit belongs to the late Norman Borlaug, who figured out how to breed crop plants that produce huge amounts of food, if you provide them with generous amounts of fertilizer, pesticide, and water.[2] But even if fertilizer and water were unlimited and pesticides did not cause pollution, there are limits to how much agricultural intensification is possible. At some point a limit to food production will be reached. Economist Kenneth Boulding said that anyone who believes that population or industrial growth can continue forever in a finite world is either a madman or an economist.[3] Futurists speculate about skyscraper greenhouses producing food for billions of people, without any assurance that we can afford to build them.[4]

Not many years ago, not only was the human population increasing but the rate of increase was increasing. That is, the upward curve was becoming steeper. Now it appears that the curve is bending the other way. The human population is still exploding. In the 1960s, there were about a hundred million more people in the world every year. Today, there are about seventy-five million more people in the world every year.[5] This is still a large number. Even if the world population continues its trajectory toward stabilization, it seems unlikely that the world will have any fewer than ten billion people in it by the end of the twenty-first century. Can the Earth support this many people?

There is no way to answer this question. It is not simply a matter of how much food we produce, but what we do with it. Agricultural production has increased so much in recent decades that nobody should be hungry, but humans have chosen to use a lot of crop production to feed livestock rather than people. Beef-eating Americans (and, increasingly, Chinese) can pay more for wheat, maize, and soybeans than can starving Somalis. Three hundred million Americans currently consume about one-third of the world's resources (energy, minerals, food, etc.). If everybody in the world had an American standard of living, the Earth could support only nine hundred million people. In fact, not even that many: the world is already consuming about 20 percent more resources than the Earth is able to regenerate. Most of the world's almost seven billion people consume much less than Americans do. Would the average American be willing to cut back consumption to the average level of humans on Earth? How many people the world can hold depends entirely on how much the world's people consume. Right now, there are not very many people actually starving to death, except in regions with political turmoil.

But current conditions cannot continue. Half of the world's people live on less than two dollars a day. The wealthiest three individuals in the world have as much collective wealth as the forty-eight poorest nations in the world. Poor people may not be starving, but the lack of adequate nutrition and access to clean water and healthcare mean that things other than starvation kill them.

Every few seconds, a child dies somewhere in the world from a disease contracted from contaminated drinking water. There are no mass famines, but poverty has plenty of other ways to kill people. As ecologist Garrett Hardin pointed out decades ago, nobody ever dies of overpopulation.[6] It is always something else that kills them.

At some point, there will be so many desperately poor people that they will not tolerate the luxury of a rich minority. What might they do? And what can we, who are rich enough to think about this problem rather than to struggle for mere survival, do to stop them? Right now, it is military power that keeps the poor in their place. At some point, the rich people and countries will not be able to afford enough military protection to continue enjoying our wealth. America can afford our military domination only because we borrow billions of dollars from other countries. Every year, there is less good farmland and easily accessible clean water, thus making the poor hungrier and more dissatisfied.

Part of the solution depends on what we do with our technology. If our resources for technological innovation are directed toward luxuries for the rich, the problem will get worse. If, on the other hand, we direct our resources toward solving the problems of the poor, then the Earth might make a relatively smooth transition into a less violent future. Helping the poor is a pretty cost-efficient thing to do. Filtration systems to provide clean drinking water are not very expensive. It costs less to help a thousand poor people to live good lives than to keep one American soldier in a war zone. Military contractors charge the government a hundred dollars to wash one fifteen-pound load of a soldier's laundry, and they will not permit the soldiers to save the government money by doing it themselves. Poor people often support terrorists. Therefore, the most cost-efficient and perhaps the only effective way to control worldwide civil unrest and terrorism is to help people rather than to use soldiers to control them. It is also the most ethical way.

The Ghost of Malthus

And here rises the ghost of Thomas Robert Malthus, the clergyman-economist of the late eighteenth century, whose work inspired both Charles Darwin and the poor laws of England. Malthus wrote that human populations (which increase exponentially) always outgrow resources (which increase, if at all, only in a linear fashion).

The human mind works in a deceptively linear fashion. We assume that tomorrow will be pretty much like today and that we can predict tomorrow by extrapolating what is happening today. But the natural world usually operates in a nonlinear fashion. The major example may be our misunderstanding of population growth. Environmental expert Lester Brown tells the story of a frog in a pond that will be completely filled with lily pads after thirty days. The lily pads double every day. On what day is the pond half filled with lily pads? The twenty-ninth day.[7] One day before the pond is completely clogged, the frog would be able to look around and think that there was still plenty of open water. The growth of human population over the last few centuries has, therefore, been a curve (a very steep one), not a line.

Overpopulation is inevitable. As explained in chapter 2, this formed the basis of Darwin's insight into natural selection. Malthus intended this as a call for moral restraint: poor people should stop having kids, and this will solve the problem. Just giving them more food will encourage them to have more kids. Malthus appears not to have intended that the British government should oppress the poor in order to depress their population, but that is what happened. Because of the poor laws, poor people were herded into workhouses that were so unpleasant they had little chance to reproduce. Of course, these laws were ineffective. But Malthus's point remains: if you give the poor food and medicine, will you not simply make the problem worse? In more recent times, this idea was expressed by Kenneth Boulding in his "utterly dismal theorem": providing help to the poor will only create more poor people, thus increasing the total sum of human misery. No haunting spirit from the past could be worse than the ghost of Malthus.[8]

Fortunately, Malthus Was Wrong

During the last decades of the twentieth century, richer nations did exactly what Malthusian economists said would make the problem worse. They provided both food and medicine to poorer nations, as direct handouts or as assistance for them to build their own economies. At the very same time, the birth rates, instead of increasing, decreased in nearly every nation in the world. Families began having fewer children because they had confidence that their children would survive. Poor people are not reproductive machines that spew out more kids when they have more food. They are intelligent persons and can decide to have fewer kids. And that is what most of them have done.

Therefore, if we use our technology to help the poor (and to prevent environmental catastrophe, which has its worst impact upon the poor), the human population might level off to a number that can, if only just, be sustained by the resources of Earth. If we use military force to maintain islands of luxury in a sea of poverty, human population may further explode.

Global Warming

The most pervasive threat to life on Earth today is global warming as a result of the greenhouse effect, and human indecision regarding what to do about it. The greenhouse effect has gotten a lot of bad press, but to a certain extent it is a good thing for the Earth. Without carbon dioxide in the atmosphere, Earth would be as cold as Mars. The climate challenge facing the people of the world today is not that the greenhouse effect is bad but that it is too much of a good thing. Having too much of a good thing can be just as disruptive as too much of a bad thing.

Among the effects of global warming are the following:

- higher temperatures
- higher sea levels, due to melting glaciers and expanding water
- more evaporation of water from oceans, causing more rainfall in coastal areas
- more evaporation of water from soils, causing droughts inland
- stronger storms
- extinctions of terrestrial species that depend on cool or humid conditions
- extinctions of marine species from acidification of ocean water
- reduced crop production in some areas
- spread of disease vectors, such as mosquitoes[9]

None of these things by itself poses an unmanageable threat, nor is the sum of the threats great enough to cause human extinction. The Earth will not become hot enough to make very many people fall over dead. The oceans will not rise fast enough that great walls of water will chase us into the hills. In the 2004 movie *The Day After Tomorrow*, global warming caused the Gulf Stream to shut down, which allowed the northern continents to suddenly freeze. This movie was intended by its creators as an exaggeration of environmental problems, in order to bring them to our attention. But even agriculture would probably not be harmed enough by the greenhouse effect to create mass starvation. If the US corn belt becomes too dry for corn to grow, then the corn can probably just be grown in Canada (though not as much, as their soil is not as good). Bad news for America, good news for Canada and Russia.

It is not human survival that is at stake, it is human civilization.

The stability of international relations and commerce depends on a stable climate. Major shifts in agriculture—if, for example, countries that currently have a lot of food such as the United States suddenly have to import food—can strain already unstable international relations. Wars could result as millions of environmental refugees flee from their flooded homes or from famine. It is the social and political disruption, which will almost certainly result from global warming, that has caught the attention of governments and businesses. The intelligence agencies of the US government consider global warming to be a threat to international stability and national security.[10] Insurance companies, and the reinsurers that insure them, are worried about the costs of storm damage. And many corporations are worried about the disruption of markets that will occur. The only corporations that seem to welcome global warming are the oil companies, which make profits when people burn fossil fuels and release carbon dioxide into the air, and the companies that can make a profit when shipping lanes open up in the Arctic Ocean. In 2009, commercial ships sailed across the Arctic Ocean for the first time in recorded history, opening the fabled "Northwest Passage" eagerly sought from the seventeenth century onward.[11]

Gaia to the Rescue?

It is possible that Gaia may save us from the greenhouse effect, as it did in the past. Photosynthesis may remove much of the carbon dioxide, which would reduce the greenhouse effect. Photosynthesis works faster, and plants grow better, when the atmosphere contains more carbon dioxide. In fact, this is one of the arguments that anti-environmentalists make against the government regulation of carbon emissions: plants will clean up the mess for us. The slogan of one such group is "CO_2—we call it life!" As an enthusiastic admirer of plants, I wish I could believe this. It's not that the statement isn't true, it's just that it will happen only over the course of many hundreds of years. Humans are pouring so much carbon dioxide into the air that plants cannot remove it quickly enough. Besides, humans are cutting down forests and paving over grasslands at the very point in time when we need their help to curb the greenhouse effect. Will plants photosynthesize enough to save the Earth from excessive warming? Not if carbon dioxide accumulation and deforestation continue at their present rates. Right now, carbon dioxide is accumulating in the atmosphere; plants have not prevented this from happening.[12]

If photosynthesis brought the greenhouse effect under control, that would be an example of the kind of negative feedback I described in chapter 2. But positive feedback processes may make the greenhouse effect worse. As carbon dioxide causes the Earth to become warmer, the higher temperatures stimulate decomposition of organic matter in peat bogs and soils. Decomposition releases yet more carbon dioxide and methane into the air, which intensifies the greenhouse effect, which intensifies decomposition, and so on. Feedback processes are one reason why major climate scientists have reported that the effects of global warming (such as rising sea levels) are, right now, occurring more rapidly than earlier predictions, based on linear extrapolations, had indicated.[13]

This is where human decisions and behavior come in. Humans are at once the most altruistic species and yet are capable of the most astonishing departures from altruism the world has ever known. Selflessness and slaughter can occur in the same place at the same time. If there are enough altruistic people, they can compensate for the evil ones. Which behavioral force will win—altruism or destruction—is impossible to predict. A small number of people with advanced technology (such as terrorists) can alter the balance of altruism and destruction—and the course of history.

The unpredictability of human behavior is one reason we cannot say exactly what will happen to the life of Earth during the next few centuries. All around the world, individuals and nations are working hard on (or at least talking hard about) reducing carbon emissions and preventing the greenhouse effect from becoming severe. There is no single solution to the problem, but the problem can be solved by making use of a lot of small solutions that already exist. We already know how to use energy more efficiently than we do. We already know how to obtain power from sunlight and wind and tides. Companies already make efficient cars and equipment for solar and wind energy. These companies are already profitable and would be more so with proper government policies.[14]

Welcome to the Republican Climate!

The world is ready to do something about the greenhouse effect—maybe not enough, but something. One of the main reasons that action has not yet been taken is that the United States has not provided enough leadership. The United States has poured more carbon dioxide into the air than any other nation and is largely responsible for the human component of global warming. China now produces more carbon dioxide than does the United States, but only barely and only recently. The failure of the United States to cooperate, much less take a leadership role, in the reduction of carbon emissions is perhaps the major reason that international cooperation on this matter has been underwhelming. And the main reason the United States has not taken decisive action is the Republican Party.

The leaders of the Republican Party have blocked nearly every policy that would allow the United States to do its part to curb global warming. There are Republican environmentalists, such as the group called Republicans for Environmental Protection, but the leaders of the Republican Party largely ignore them.[15] Most Republican leaders are even unwilling to admit that the problem exists.[16] Oklahoma senator Jim Inhofe even suggests that global warming is a hoax created by people like Al Gore and me just so we can sell books. (Let me point out that the CEO of Exxon Mobil earns in a few seconds what I will earn from this book or from my earlier book that dealt extensively with this issue.) Former president George W. Bush even thought it was a joke. When he left a 2008 meeting of the world's leading economic powers, he said, "Good-bye from the world's biggest polluter."[17]

The motivation of the leaders of the Republican Party to oppose action against the greenhouse effect is a short-term focus on money. Republican politicians benefit from campaign contributions provided by oil and coal companies. The Supreme Court decision in January 2010 to allow unlimited corporate campaign spending has buried all hope that politicians might act in the public interest by taking steps to control such things as the greenhouse effect.

There are, in fact, plenty of businesses that want to see the greenhouse effect prevented. Many of them have begun making large investments in technology (such as wind and solar energy or fuel-efficient cars) that will help to prevent the greenhouse effect.[18] But the greenhouse effect will have its major impact decades from now and will cost us a lot of money at that time. Meanwhile, this year, this decade, it is cheaper to keep pouring the carbon into the air. Democratic lawmakers, many of whom are also recipients of coal and oil money, were at first eager to bring carbon emissions under control, but have recently abandoned the issue. Republican lawmakers usually vote as a bloc and now present an invincible barrier to further action on climate change.

How can Republican leaders deny the greenhouse effect? The answer to this question takes us into strange territory. It appears to me that they believe very strongly in the Gaia Hypothesis, though they will not admit it. They believe that the Earth can neutralize any abuses that we put upon it. Plants will soak up all the extra carbon. Don't worry about carbon dioxide—Earth will take care of it. Relax and make money. They also take this approach to all forms of pollution. Don't worry about pesticides—the Earth will break them down. Don't worry about mercury coming from the smokestacks of coal-fired power plants—nature will clean it up somehow. It is almost as if they consider Gaia to be a real goddess, rather than just an imperfect system of negative feedback. They appear willing to place the survival of human civilization on the altar of this goddess. This is particularly strange, since the Republican Party is the most outspoken source of support for Christian theology. They condemn liberal earth-goddess environmentalism, while at the same time acting as if they believe in it.[19]

As noted previously, the Gaia network of life is much better at regulating the conditions of Earth than it was in the distant past, but this ability is limited. Human activities are pushing Earth beyond the ability of Gaia to regulate it, at least well enough for humans to live comfortably in.[20] The result is likely to be an Earth that is much warmer—not warm enough to threaten life in general but warm enough to be disastrous to human activities.

If the Republicans succeed in blocking progress in dealing with the greenhouse effect, the next few centuries may be severely disruptive for humankind. The greenhouse effect won't kill us, but it may bankrupt us. What do we do about the major cities that will be flooded by rising ocean levels? Whether we build big sea walls (oversized versions of Dutch dikes) or whether we relocate all the people and infrastructure, the cost will be trillions of dollars worldwide. Famines in some areas will be incompletely compensated by food production in other areas, and the free market will cause food prices to soar. Wars over resources, even over places to live, appear inevitable. All this can happen even if the temperature increase is as little as five degrees C (nine degrees F).

The people living in the greenhouse future (and there will still be plenty of them) may look back and curse us for making their world so catastrophic. They will rightly say that we knew what we were doing and that we failed to take actions that we knew we should take. And they will blame the Republican Party. They will not remember that some Republicans were trying to take their party in a more responsible direction. I predict that they will refer to their new world as the "Republican Climate." They will pass judgment on us the same way we look back on the slave owners of the Confederacy and condemn them. We criticize Thomas Jefferson for failing to free his slaves, even though this was his announced intention. If you were in Thomas Jefferson's shoes, you might find yourself unable to free your slaves. The whole economy of the South, including Jeffersonian Virginia, was based on slavery. And it was illegal to free a slave unless you could pay for him or her to have a subsistence-level job afterward. Jefferson, apparently, could not afford to do this. Most modern observers have little sympathy for these problems. And the people of the future will have little sympathy for us when we claim that trying to fight the greenhouse effect might make some consumer items more expensive or cause our taxes to rise.

THE DISTANT HUMAN FUTURE

What will life on Earth look like beyond the immediate future of a greenhouse world? Uncertainties mount, and wildly different scenarios are possible. But all of them, short of the unlikely event of human extinction, involve a great deal of ecological disturbance.

Ecological disturbances such as fires and storms are natural, but humans have caused natural disturbances to be more frequent and intense, and have created new kinds of disturbances. This started with the extensive transformation of natural landscapes into agricultural lands and cities. One consistent feature of ecological disturbance is that the species that flourish in these places are those that grow rapidly and produce a lot of offspring—that is, weeds and vermin. It is therefore safe to predict that the dominant plants and animals of a future Earth will be the evolutionary descendants of today's pests.[21] Largely absent will be the big animals that reproduce slowly and the trees that live for thousands of years. In their place will be futuristic rats and pigs and tree-sized descendants of weeds. Fast-growing trees, such as tree-of-heaven and mulberry, will probably be an important part of the landscape. Today, these trees live in places where forests have been recently destroyed; trees-of-heaven are particularly common in abandoned industrial sites. There will be forests on the future Earth, but probably not of oaks, maples, or giant sequoias (figure 9-1).

Figure 9-1. The General Sherman tree, living in Sequoia National Park in the Sierra Nevada of California, is the largest organism on Earth. It is over two thousand years old. Future forests, disturbed by humans, are unlikely to contain such large and long-lived trees.

The change will be gradual enough that future humans may not recognize the difference. Today, when people hike through the deciduous forests of eastern North America, almost the only thing they see is second-growth forests, which have filled in the landscape after the primordial forests were cut down in the previous century. Large trees, with trunks one to several yards in diameter, were common before European and American settlers began cutting them down. Yet today, hikers do not usually comment on the absence of huge trees, which, of course, they have never seen. Most hikers in the Appalachians are unaware that only a few decades ago chestnut trees dominated these forests and then were all killed (except for scattered rootstocks that continue to sprout) by a fungus introduced from Europe. There will probably be hikers and campers in the future—a love of nature appears to be an ineradicable part of the human mind—and when they see lush green forests of mulberry and tree-of-heaven, they will not think about the absence of oak and maple forests that they have never seen. Our photographs and descriptions of the magnificent forests and elephant herds of today will seem as unreal to them as the huge trees painted by the artists of the Hudson River School in early nineteenth-century America seem to us. A few primordial forests remain today but are not widely known. Białowieża Puszcza is a primordial forest on the border of Poland and Belarus (where it is called Belovezhskaya Pushcha). This forest, with huge trees, is pretty much the way all of Europe used to be, but few people see and visit it. A few remnant oak and maple forests may persist, largely unnoticed, into future centuries the way Białowieża survives today.[22]

How much natural area will survive is an open question. If human society becomes and remains disorganized, humans may be spread out over large areas, each with extensive ecological disturbance and filled with weeds and pests. But if human society becomes more organized, most humans may be able to live in technologically sophisticated cities, allowing much of the landscape to heal, magnificent forests to rebound from the tiny Białowieża-like remnants, and wildlife to return from isolated herds.

One of the hardest things to predict is the future of technology. I, at least, consistently underestimate technological breakthroughs. I wrote a novel manuscript in 1990, set in the year 2075, that included film cameras and cassette tape recorders, both of which are already rare. At the same time, technology is a fragile thing. Modern technology depends entirely on the persistence of a stable civilization. If civilization faltered, even without being destroyed, there might be no memory of how to access information stored in computers; much of our art and science may be lost. Our future may resemble the one novelist and Nobelist Doris Lessing created in *Mara and Dann*, in which there were still trains and airplanes, but they had to be pushed.

Then there is the future evolution of the human species itself. In prehistoric times, humans diversified into several races and numerous ethnicities, isolated from one another. There were some prehistoric comminglings, such as when Indonesians and Africans mixed to become the modern Malagasy of Madagascar. Today, the human species appears headed toward a homogenization. A disruption of travel and commerce could change all of this, setting up reproductive barriers that would most likely promote the evolution of new and separate races, not necessarily the preservation of current races. This was another theme in *Mara and Dann*.

No writer has surpassed the insight of Herbert George Wells in his novel *The Time Machine*, in which humans have evolved into an effete, privileged, very weak upper class (the Eloi) and a brutish lower class (Morlocks). The Eloi, living in the sunlight, have degenerated into childlike stupidity, not realizing that they are, in effect, cattle for the tunnel-dwelling Morlocks to eat. This very thing has happened in some species of ants, in which the ruler ants are helpless, taken care of by strong slave ants that are very much in control of the interaction. In the hundred years since Wells wrote this novel, a scenario like this has remained conceivable, though unlikely, for the future of humankind.

There are few things we can predict for certain about human evolution, especially if we begin to control our own evolution through genetic engineering.[23] As explained in this book, human intelligence is limited by brain size, and our brains are as big as they can be for a species whose babies are born through a birth canal from a uterus. But is it possible that technology will allow babies to be born after just three months' gestation? Their brains could continue to grow to a large size after birth, if they have the genetic basis for it. Or perhaps cesarean sections will become the norm? Should this occur, human brain size is limited only by the ability of the neck to hold up the head—and maybe not even that. But these things may occur only if technology remains uninterrupted.

The most tragic thing that could happen to our species, and to the Earth that is so greatly impacted by it, would be the loss of altruism. Feelings of altruism and of selfishness are both innate, but the ways of putting these feelings to work must be learned. Young apes learn the direct reciprocity of grooming by watching older apes. Young humans feel hatred, always, but directing it toward certain racial or ethnic targets is something they have to learn. As Oscar Hammerstein II wrote in the musical *South Pacific*, "You've got to be taught before it's too late / Before you are six or seven or eight / To hate all the people your relatives hate . . ." Sufficient disruption of human society might leave mere tatters of altruism, and select for those individuals who feel the least of it. Altruism, that crowning achievement of evolution, of Gaia, could be almost lost, and with it will be lost any sense of worldwide civilization, without which humans have no chance of adapting to the inevitable changes that will take place on Earth.

Or not. No one can predict when or where genius individuals may come along who may have an impact unimaginably greater than their humble lives might suggest. How many millions of lives have been made more altruistic by the words and the examples of Siddhartha Gautama, later called Buddha, and Jesus of Nazareth, later called Christ? Under the guidance of some such person, human society may gather up the shards of altruism and reassemble it into a healthy civilization.

THE END

Whether humankind develops a utopia or a dystopia, or anything in between, the time will come when Earth will become uninhabitable.

A Human Future in Space?

Many people speculate that the ultimate human destiny is to spread into outer space. Physicist Frank Tipler goes so far as to claim that human expansion throughout the universe is not only inevitable but will quite literally be immortality.[24]

But human expansion into space is such a large project that it is nearly inconceivable. It certainly is no way for us to try to escape from the messes we have created on Earth. Human expansion into space, aside from a few manned and a lot of unmanned missions, will require the kind of international cooperation that we have never been able to achieve. The idea that biodiversity can be saved by putting it all aboard a spaceship (think of the 1972 movie *Silent Running*) is an interesting story and nothing more.

Human migration into outer space is extremely unlikely to occur because it would be so fantastically expensive. Trillions of dollars would be required to send enough people and equipment to the moon to establish a colony. It is true that some of the costs could be reduced if hydrogen fuel and oxygen could be generated from the solar-powered splitting of water molecules that are now known to exist in lunar soil—but only if the water exists at high enough concentration that it would not take more energy to extract it than could be generated from it. It would cost even more, by several orders of magnitude, to establish a Mars colony. Plans to alter the climate of Mars in such a way as to make it more suitable for human habitation would be nearly inconceivable, despite the scenarios popularly presented on science-oriented cable channels. Mars could never be "terra-formed" such that people could walk around outside in sweaters and breathe the air.[25]

One reason space migration would be so expensive is the distances involved. They are so great that even the minds of space enthusiasts cannot process them. The moon is not very far away, but months of weightlessness on a journey to Mars might be debilitating to a human body no matter what compensatory therapies may be provided in the spaceship. And the orbit of Mars is only the small inner disc of the solar system. It is almost as far from Mars to Jupiter as from Mars to the sun. It took a Voyager spacecraft over thirty years to travel from Earth to the heliopause, which is the edge of influence of the sun's particles. A small piece of machinery might conceivably be propelled to a speed close to that of light by a large antenna collecting solar particles, which means that it could reach Proxima Centauri a couple of years after accelerating to that speed—but nothing as large as a human could benefit from that manner of propulsion. A long journey would exceed the capacity of a spaceship to carry enough food for a person. Passengers could raise their food in an onboard ecosystem (which would have just one chance to work right), and then only when they are close enough to a star to use its energy.

A human future in space is the only way to avoid extinction when, five billion years from now, the sun becomes large enough that its radiation will kill us. But even if we had millions of years to work on it, would humans cooperate with one another enough, tap the resources of altruism enough to complete the job?[26] Every century or so, we could perhaps send a few more items to Mars so that, in maybe three thousand years, a colony could be assembled. That would be plenty of time to get ready for the sun's red bloating, which would sterilize Earth, but also might, for a while, make Mars a tropical paradise. Or not. The smile of Carl Sagan, pondering Fermi's Paradox, returns to remind us that advanced civilizations have apparently not spread through our part of the galaxy in such a manner.

Therefore, it is unlikely that humankind can outlive Gaia; in fact, humanity is unlikely to even see Gaia into old age.

The End of Earth

As mentioned previously, the sun will begin to expand, becoming a red giant in about five billion years from now. When the sun reaches its final size, its surface will be farther from its center than the orbit of Earth is now. That is, Earth will be inside the sun's mass. Even if Earth is not vaporized, all of the water, atmosphere, and life upon it will be. Long before this happens, the intensifying wind of solar particles, from which the magnetic field currently protects us, will take away enough water and atmosphere that only bacteria will be able to survive (or the last humans, in underground bunkers). The negative feedback systems of Gaia will be overwhelmed perhaps only a billion years from now, about the time our car, having left Kansas, reaches Grand Junction. Near the end, Gaia will be a sparse network of green scum and stromatolites, in tepid seas, as it was three billion years ago. Its air might still have oxygen, produced by the scum, and it might be breathable by animals—a moot point in a world so hostile to complex life-forms. But this is at least one thing in which we can place our faith: until the last minute that life is possible on the surface of Planet Earth, there will be photosynthesis. There will be no final mourning of Gaia's death, which will be experienced by only a few cyanobacteria on the surface and mineral-eating bacteria deep in the rocks.

Also about five billion years from now, the Milky Way galaxy will "collide" with the Andromeda galaxy. It will probably not be much of a collision, since both galaxies consist of extremely widely spaced stars except at their centers. An Andromedan star might pass close enough to the solar system to destroy it—or not.

The End of the Universe

If, after about ten billion years, there is a human colony on one of the moons of Jupiter, its members would look out into the sky and see the billions of stars of our Milky Way galaxy. They may not, however, be able to see other galaxies, which may then have receded, one by one, beyond the limits of visibility, except for the Andromeda galaxy, which will have merged with ours. Astrophysicists have calculated an inflated expansion rate for galaxies, which implies that expansion will continue forever (chapter 1).

Galaxies are getting farther apart, but stars within galaxies are not. There will be stars in the sky as long as stars in our galaxy burn. Some of them are already burning out; others are still being born by condensation of gases and dust inside nebulae. The last stars will not burn out until one hundred trillion years from now. In this far distant future, the sky will be black and the universe will be dark. What happens after the death of Earth and of the universe is as unknowable as what happens after human death.

BUT BEFORE THEN . . .

I end this book with a description of what we know we must do on the only timescale that is meaningful to us. Early in the book, I explained how photosynthesis was one of the most essential processes within the life of Gaia. While photosynthesis will exist as long as Gaia does, we need to make sure we continue to have enough of it. And there are lots of other things that trees and other land plants do besides photosynthesis that are essential to maintaining conditions suitable for human survival.

We humans think that we are the most important things in the world. How could the world operate without us? But a world could operate just fine not only without humans but without any animals at all. All it would need is plants, which carry out photosynthesis, and decomposers, which recycle the carbon through their respiration.

About half of the photosynthesis in the world is carried out by microbes and seaweeds in the oceans. The most abundant cell in the world is *Prochlorococcus*, a tiny marine photosynthetic bacterium, which was not discovered until 1988. About half of the photosynthesis in the world is carried out by plants on land, even though the oceans cover a much larger portion of Earth's surface than does dry land. The open oceans are relatively poor in the nutrients that photosynthetic organisms need. Most marine photosynthesis occurs in waters near the continents, which are renewed by minerals discharged into them by rivers or brought up from the ocean depths by currents. On the continents, especially in warm regions with plenty of rain, trees and other plants carry out a lot of photosynthesis.

Plants do a lot more than just photosynthesis. They are the guts of Gaia and do a lot of other things that are necessary functions of Gaia. In the first chapter of Genesis, an ancient poem about the beauty, order, and abundance of creation, God created plants on the same day as the dry land. This poetically affirms that a landscape without plants is incomplete and dysfunctional. Among the things that trees and other plants do are the following:

- They regulate the flow and cycling of water. When rain falls on a bare hillside, the water rushes down the slope and floods the land and people below. Then when the rain stops, there is no more water in the ground than there was before. Trees, shrubs, and grasses slow down the impact of the raindrops, allowing the rain to penetrate deeply into the soil and renew the underground water supplies. By doing this, the plants prevent floods. This fact has been experimentally demonstrated by comparing forested hillsides with adjacent hillsides on which the trees have been cut down: the clear-cut hillsides have floods.[27] Plants absorb much of the water for their own use, which evaporates from their leaves (a process called transpiration) into the air. Some of this water vapor becomes clouds. Therefore, a forest can create clouds. About half of the rain that falls on the Amazon rain forest comes from water that evaporated from the rain forest trees.[28]
- They prevent soil erosion and mudslides. Wind and rain cannot easily dislodge soil that is protected by grasses, shrubs, or trees.
- They produce all the food in the world. Photosynthesis does not just produce oxygen and consume carbon dioxide; it also uses the carbon dioxide and energy from the sun to create food. They are the basis of the food chain.

Imagine a machine that could solve all our problems. It produces oxygen for us to breathe, food for us to eat, medicines and many other products we need in our lives. It protects us from floods and droughts. It keeps us from overheating by removing carbon dioxide from the air. To build and operate a billion such machines would produce more carbon dioxide than the machines would remove from the air, unless we could get the machines to build themselves and run on solar energy. Could such wonderful machines ever exist? They do—and we call them plants. We do not have to make them do it or even thank them, we only have to allow them. And they do it in complete silence and unutterable beauty.[29] This is part of the fundamental work of Gaia. Plants do these things for themselves, but we are among the beneficiaries. The "ecosystem services" they provide have been estimated to be worth $33 trillion in 1997 dollars.[30]

It has long been recognized, and as long ignored, that plants have to be allowed to do their work, or else we will not be able to survive. Consider one example: the ancient kingdom of Judah, as interpreted by the writers of the Old Testament books of the Chronicles. When the kingdom of Judah was at last conquered by Babylon, the chronicler attributed it to the sins of the people. But there is one particular sin the chronicler singled out. One of the many Old Testament commandments was to allow agricultural land to lie fallow every seventh year. This "Sabbath of the fields," we now understand, would have allowed wild plants to partially restore the fertility of the soil and diseases and pests specific to the crops to die away. To the ancient Israelites, it was perhaps solely understood in terms of religious service, but the Sabbath of the fields was apparently never put into practice. Thus, the chronicler implied, the people of Judah owed God one-seventh of the years that their kingdom had existed. According to tradition, the Jewish kingdom existed for 490 years, so the people therefore owed the land and its God seventy years of rest. They did not render this offering while the kingdom stood, so the land rested during the seventy years of Israelite captivity in Babylon. Among the very last words of the second book of Chronicles is this chilling statement: "And so the land enjoyed its rests. All the days that it lay desolate it kept Sabbath, to fulfill seventy years." The implication, as much for us as for those ancient people, is that one way or another, the land will get its rest, its chance to recover. We do not need to stop our economic activities in order to allow this to happen. We can incorporate a modern equivalent of the Sabbath of the fields into our economic system by replanting forests or by using a form of agriculture that does not harm the soil. In this way, the land can rest the way the heart rests—between each contraction, without missing a beat. But we must either incorporate stewardship of the land into our economic system, allowing plants to renew the land on a regular basis, or else the land will enjoy its Sabbaths after the collapse of our economic system. Now, or later, the land will rest.[31] This is the choice that Gaia demands of us.

NOTES

INTRODUCTION

1. Mark Fonstad, William Pugatch, and Brandon Vogt, "Kansas Is Flatter Than a Pancake," *Journal of Improbable Research* 9 (2003): 16–17. Available online at http://improbable.com/airchives/paperair/volume9/v9i3/kansas.html (accessed February 8, 2010).

2. Sean Carroll, *From Eternity to Here: The Quest for the Ultimate Theory of Time* (New York: Dutton, 2010); Neil deGrasse Tyson and Donald Goldsmith, *Origins: Fourteen Billion Years of Cosmic Evolution* (New York: Norton, 2004).

3. Bill Bryson, *A Short History of Nearly Everything* (New York: Random House, 2003), pp. 127–32.

4. Ibid., pp. 11–12.

5. Erick T. Young, "Cloudy with a Chance of Stars," *Scientific American*, February 2010, pp. 34–41.

6. Martin Bojowald, "Follow the Bouncing Universe," *Scientific American*, October 2008, pp. 44–51.

7. David C. Catling and Kevin J. Zahnle, "The Planetary Air Leak," *Scientific American*, May 2009, pp. 36–43.

8. Owen Gingerich, *The Book Nobody Read: Chasing the Revolutions of Nicolaus Copernicus* (New York: Walker, 2004).

9. Marshall Hall, "The Non-Moving Earth and Anti-Evolution Web Page of the Fair Education Foundation, Inc," accessed February 8, 2010, http://www.fixedearth.com.

10. Steven Dutch, "21st Century Geocentrism," accessed February 8, 2010, http://www.uwgb.edu/dutchs/PSEUDOSC/Geocentrism.htm.

CHAPTER 1

1. G. Evelyn Hutchinson, *The Ecological Theater and the Evolutionary Play* (New Haven, CT: Yale University Press, 1965).

2. Stephen H. Schneider et al., eds., *Scientists Debate Gaia: The Next Century* (Cambridge, MA: MIT Press, 2004).

3. Peter Ward, *The Medea Hypothesis: Is Life on Earth Ultimately Self-Destructive?* (Princeton, NJ: Princeton University Press, 2009).

4. "PlanetQuest: Exoplanet Exploration," Jet Propulsion Laboratory, National Aeronautics and Space Administration, accessed February 10, 2010, http://planet quest.jpl.nasa.gov; Michael W. Werner and Michael A. Jura, "Improbable Planets," *Scientific American*, June 2009, pp. 38–45.

5. Paul Kalas et al., "Optical Images of an Exosolar Planet 25 Light-Years from Earth," *Science* 322 (2008): 1345–48.

6. Peter D. Ward and Donald Brownlee, *Rare Earth: Why Complex Life Is Uncommon in the Universe* (New York: Copernicus, 2000).

7. Satoshi Mayama et al., "Direct Imaging of Bridged Twin Protoplanetary Disks in a Young Multiple Star," *Science* 327 (2010): 306–308.

8. Walter Alvarez, *T. Rex and the Crater of Doom* (Princeton, NJ: Princeton University Press, 1997).

9. Yaoming Hu et al., "Large Mesozoic Mammals Fed on Young Dinosaurs," *Nature* 433 (2005): 149–52.

10. Michael J. Benton, *When Life Nearly Died: The Greatest Mass Extinction of All Time* (London: Thames and Hudson, 2003); Douglas H. Erwin, *Extinction: How Life on Earth Nearly Ended 250 Million Years Ago* (Princeton, NJ: Princeton University Press, 2006).

11. Stephen Jay Gould, "The Cosmic Dance of Shiva," in *The Flamingo's Smile: Reflections in Natural History* (New York: Norton, 1985), p. 445.

12. Surenda Verma, *The Tunguska Fireball* (London: Icon, 2004).

13. Robert M. Hazen, *Genesis: The Scientific Quest for Life's Origin* (Washington, DC: Joseph Henry Press, 2005); James Trefil, Harold J. Morowitz, and Eric Smith, "The Origin of Life," *American Scientist* 97 (2009): 206–13; Arturo Becerra et al., "The Very Early Stages of Biological Evolution and the Nature of the Last Common Ancestor of the Three Major Cell Domains," *Annual Review of Ecology, Evolution, and Systematics* 38 (2007): 361–79.

14. Alonso Ricardo and Jack W. Szostak, "Life on Earth," *Scientific American*, September 2009, pp. 54–61.

15. James E. Lovelock, *Gaia: A New Look at Life on Earth* (New York: Oxford University Press, 1987); Lynn Margulis, *Symbiotic Planet: A New Look at Evolution* (New York: Basic Books, 1998).

16. Steven Johnson, *Emergence: The Connected Lives of Ants, Brains, Cities, and Software* (New York: Scribner, 2001).

17. David Beerling, *The Emerald Planet: How Plants Changed Earth's History* (Oxford: Oxford University Press, 2007).

18. Raymond B. Huey and Peter D. Ward, "Hypoxia, Global Warming, and Terrestrial Late Permian Extinctions," *Science* 308 (2005): 398–401.

19. Tyler Volk, "Gaia Is Life in a Wasteworld of By-Products," in Schneider et al., *Scientists Debate Gaia*, pp. 27–36.

20. Beerling, *Emerald Planet.*

21. Benton, *When Life Nearly Died.*

22. Gabrielle Walker, *Snowball Earth: The Story of a Great Global Catastrophe That Spawned Life As We Know It* (New York: Crown Publishers, 2001); Andrew H. Knoll, *Life on a Young Planet: The First Three Billion Years of Evolution on Earth* (Princeton, NJ: Princeton University Press, 2003), pp. 211–15.

23. Mark A. S. McMenamin, "Gaia and Glaciation: Lipalian (Vendian) Environmental Crisis," in Schneider et al., *Scientists Debate Gaia*, pp. 115–28.

24. Iosif S. Shklovsky and Carl Sagan, *Intelligent Life in the Universe* (San Francisco: Holden-Day, 1966).

25. Richard P. Turco, Owen B. Toon, T. P. Ackerman, J. B. Pollack, and Carl Sagan, "Nuclear Winter: Global Consequences of Multiple Nuclear Explosions," *Science* 222 (1983): 1283–92.

26. Owen B. Toon, Alan Robock, Richard P. Turco, Charles Bardeen, Luke Oman, and Georgiy L. Stenchikov, "Consequences of Regional-Scale Nuclear Conflicts," *Science* 315 (2007): 1224–25.

27. Alice Calaprice, *The New Quotable Einstein* (Princeton, NJ: Princeton University Press, 2005), p. 173.

CHAPTER 2

1. Carl Woese, *The Genetic Code* (New York: Harper & Row, 1968); Walter Gilbert, "The RNA World," *Nature* 319 (1986): 618.

2. Richard Dawkins, *The Blind Watchmaker*, chap. 3 (New York: Oxford University Press, 1986).

3. Daniel E. Dykhuizen, "Experimental Studies of Natural Selection in Bacteria," *Annual Review of Ecology and Systematics* 21 (1990): 373–98; J. E. Barrick et al., "Genome Evolution and Adaptation in a Long-Term Experiment with *Escherichia coli*," *Nature* 461 (2009): 1243–47; Richard Lenski, "Experimental Evolution," accessed February 13, 2010, https://myxo.css.msu.edu/ecoli/publist.html.

4. Stanley A. Rice, "Agriculture, Evolution of," and "Artificial Selection," in *Encyclopedia of Evolution* (New York: Facts on File, 2006).

5. Stanley A. Rice, "Resistance, Evolution of," in *Encyclopedia of Evolution*; Michael Schnayerson and Mark J. Plotkin, *The Killers Within: The Deadly Rise of Drug-Resistant Bacteria* (New York: Back Bay Books, 2003); "Antibiotic/Antimicrobial Resistance," Centers for Disease Control and Prevention, accessed February 13, 2010, http://www.cdc.gov/drugresistance/index.htm.

6. Rachel Carson, *Silent Spring*, chap. 16 (New York: Houghton Mifflin, 1962).

7. David Lack, *Darwin's Finches* (Cambridge: Cambridge University Press, 1947).

8. Jonathan Weiner, *The Beak of the Finch: A Story of Evolution in Our Time* (New York: Knopf, 1994); Peter R. Grant and B. Rosemary Grant, *How and Why Species Multiply: The Radiation of Darwin's Finches* (Princeton, NJ: Princeton University Press, 2007).

9. Niles Eldredge, *The Pattern of Evolution* (New York: W. H. Freeman, 1999).

10. Lynn Margulis, *Symbiotic Planet: A New Look at Evolution* (New York: Basic Books, 1998).

11. Lynn White Jr., "The Historical Roots of Our Ecologic Crisis," *Science* 155 (1967): 1203–1207.

12. Stanley A. Rice, *Green Planet: How Plants Keep the Earth Alive* (New Brunswick, NJ: Rutgers University Press, 2009).

13. Michael Pollan, *The Botany of Desire: A Plant's-Eye View of the World* (New York: Random House, 2001).

14. Mark A. S. McMenamin, *The Garden of Ediacara: Discovering the First Complex Life* (New York: Columbia University Press, 1998).

15. Becky L. Williams, Edmund D. Brodie Jr., and Edmund D. Brodie III, "A Resistant Predator and Its Toxic Prey: Persistence of Newt Toxin Leads to Poisonous (Not Venomous) Snakes," *Journal of Chemical Ecology* 30 (2004): 1901–19.

16. Richard Dawkins, *The Selfish Gene*, chap. 11 (New York: Oxford University Press, 1976).

17. Alex Thornton and Katherine McAuliffe, "Teaching in Wild Meerkats," *Science* 313 (2006): 227–29.

18. Daniel J. Levitin, *This Is Your Brain on Music: The Science of a Human Obsession* (New York: Penguin, 2007).

19. Stephen Jay Gould, *Full House: The Spread of Excellence from Plato to Darwin* (New York: Harmony Books, 1996).

CHAPTER 3

1. Mikhail V. Matz et al., "Giant Deep-Sea Protist Produces Bilaterian-Like Traces," *Current Biology* 18 (2008): 1–6.

2. Mark A. S. McMenamin, *The Garden of Ediacara: Discovering the First Complex Life* (New York: Columbia University Press, 1998).

3. Stephen Jay Gould, *Wonderful Life: The Burgess Shale and the Nature of History* (New York: Norton, 1989); Simon Conway Morris, *The Crucible of Creation: The Burgess Shale and the Rise of Animals* (New York: Oxford University Press, 1998).

4. Sam Gon III, "The *Anomalocaris* Homepage," accessed February 13, 2010, http://www.trilobites.info/anohome.html.

5. John A. Long, *The Rise of the Fishes: 500 Million Years of Evolution* (Baltimore, MD: Johns Hopkins University Press, 1996).

6. Linda E. Graham, Martha E. Cook, David T. Hanson, Kathleen B. Pigg, and James M. Graham, "Structural, Physiological, and Stable Carbon Isotopic Evidence That the Enigmatic Paleozoic Fossil *Prototaxites* Formed from Rolled Liverwort Mats," *American Journal of Botany* 97 (2010): 268–75.

7. Mark A. S. McMenamin and Dianna L. S. McMenamin, *Hypersea* (New York: Columbia University Press, 1996).

8. Neil Shubin, *Your Inner Fish: A Journey into the 3.5-Billion-Year History of the Human Body* (New York: Pantheon, 2008).

9. Quanguo Li et al., "Plumage Color Patterns of an Extinct Dinosaur," *Science Express*, published online February 5, 2010, accessed February 13, 2010, http://www.sciencemag.org/cgi/rapidpdf/science.1186290v1.pdf.

10. Xiao-Ting Zheng et al., "An Early Cretaceous Heterodontosaurid Dinosaur with Filamentous Integumentary Structures," *Nature* 458 (2009): 333–36.

11. Ian Tattersall and Jeffrey Schwartz, *Extinct Humans* (New York: Basic Books, 2001).

12. Steven M. Stanley, *Children of the Ice Age: How a Global Catastrophe Allowed Humans to Evolve* (New York: Harmony Books, 1996).

13. Tim D. White et al., "Macrovertebrate Paleontology and the Pliocene Habitat of *Ardipithecus ramidus*," *Science* 326 (2009): 87–93; C. Owen Lovejoy et al., "The Great Divides: *Ardipithecus ramidus* Reveals the Postcrania of Our Last Common Ancestors with African Apes," *Science* 326 (2009): 100–106.

14. Naama Goren-Inbar et al., "Evidence of Hominin Control of Fire at Gesher Benot Ya'aqov, Israel," *Science* 304 (2004): 725–27.

15. Richard Wrangham, *Catching Fire: How Cooking Made Us Human* (New York: Basic Books, 2009).

CHAPTER 4

1. Mitch Leslie, "On the Origin of Photosynthesis," *Science* 323 (2009): 1286–87.

2. Lynn Margulis and Dorion Sagan, *Acquiring Genomes: A Theory of the Origins of Species* (New York: Basic Books, 2002); Andrew H. Knoll, *Life on a Young Planet: The First Three Billion Years of Evolution on Earth* (Princeton, NJ: Princeton University Press, 2003), pp. 122–38.

3. Kwang W. Jeon, "Integration of Bacterial Endosymbionts in Amoebae," *International Review of Cytology* 14 (1983): 29–47.

4. Noriko Okamoto and Isao Inouye, "A Secondary Symbiosis in Progress?" *Science* 310 (2005): 287.

5. Xunlai Yuan, Shuhai Xiao, and Thomas N. Taylor, "Lichen-Like Symbiosis 600 Million Years Ago," *Science* 308 (2005): 1017–1020.

6. Winfried Remy, Thomas N. Taylor, Hagen Hass, and Hans Kerp, "Four-Hundred-Million-Year-Old Vesicular Arbuscular Mycorrhizae," *Proceedings of the National Academy of Sciences USA* 91 (1994): 11841–43.

7. Jennifer M. Robinson, "Lignin, Land Plants, and Fungi: Biological Evolution Affecting Phanerozoic Oxygen Balance," *Geology* 15 (1990): 607–10.

8. Dong Ren et al., "A Probable Pollination Mode before Angiosperms: Eurasian, Long-Proboscid Scorpionflies," *Science* 326 (2009): 840–47.

9. William C. Burger, *Flowers: How They Changed the World* (Amherst, NY: Prometheus Books, 2006).

10. Danny Kessler, Klaus Gase, and Ian T. Baldwin, "Field Experiments with Transformed Plants Reveal the Sense of Floral Scents," *Science* 321 (2008): 1200–1202.

11. Victor Rico-Gray and Paolo S. Oliveira, *The Ecology and Evolution of Ant-Plant Interactions* (Chicago: University of Chicago Press, 2007); Todd M. Palmer et al., "Breakdown of an Ant-Plant Mutualism Follows the Loss of Large Herbivores from an African Savanna," *Science* 319 (2008): 192–95, summarized by Mitch Leslie, "The Importance of Being Eaten," *Science* 319 (2008): 146–47.

12. Andrew P. Norton et al., "Mycophagous Mites and Foliar Pathogens: Leaf Domatia Regulate Tritrophic Interactions in Grapes," *Ecology* 81 (2000): 490–99.

13. Elizabeth A. Grice et al., "Topographical and Temporal Diversity of the Human Skin Microbiome," *Science* 324 (2009): 1190–92, summarized by Christopher J. Marx, "Getting in Touch with Your Friends," *Science* 324 (2009): 1150–51; J. Xu and J. I. Gordon, "Honor Thy Symbionts," *Proceedings of the National Academy of Sciences USA* 100 (2003): 10452–59.

CHAPTER 5

1. James Thurber and E. B. White, *Is Sex Necessary?: Or Why You Feel the Way You Do*, 75th anniversary ed. (New York: Harper, 2004).

2. Charles Darwin, *On the Origin of Species by Means of Natural Selection*, 150th anniversary ed. (New York: Signet, 2003); Charles Darwin, *The Descent of Man, and Selection in Relation to Sex* (New York: Penguin, 2004). Avoid the edition with an introduction by creationist Ray Comfort. Comfort's introduction accuses Darwin of racism and dishonesty, and contains a brief biographical summary that was plagiarized from another author.

3. Leigh Van Valen, "A New Evolutionary Law," *Evolutionary Theory* 1 (1973): 1–30.

4. Christopher G. Wilson and Paul W. Sherman, "Anciently Asexual Bdelloid Rotifers Escape Lethal Fungal Parasites by Drying Up and Blowing Away," *Science* 327 (2010): 574–76.

5. Ashley Montagu, *The Natural Superiority of Women*, 5th ed. (Lanham, MD: Rowman & Littlefield, 1999).

6. Stuart Bearhop et al., "Assortative Mating as a Mechanism for Rapid Evolution of a Migratory Divide," *Science* 310 (2005): 502–504.

7. Matt Ridley, *The Red Queen: Sex and the Evolution of Human Nature*, chap. 7 (New York: Harper, 2003); Olivia Judson, *Dr. Tatiana's Sex Advice to All Creation* (New York: Henry Holt, 2002); Lynn Margulis and Dorion Sagan, *Mystery Dance: On the Evolution of Human Sexuality* (New York: Simon & Schuster, 1991); Steve Jones, *Y: The Descent of Men* (Boston: Houghton Mifflin, 2003).

8. Jonathan Weiner, *The Beak of the Finch: A Story of Evolution in Our Time* (New York: Knopf, 1994).

9. Malte Andersson, "Female Choice Selects for Extreme Tail Length in a Widowbird," *Nature* 299 (1982): 818–20.

10. Mary F. Willson and Nancy Burley, *Mate Choice in Plants: Tactics, Mechanisms, and Consequences* (Princeton, NJ: Princeton University Press, 1983).

11. Tao Zhang and Dun-Yan Tan, "An Examination of the Function of Male Flowers in an Andromonoecious Shrub *Capparis spinosa*," *Journal of Integrative Biology* 51 (2009): 316–24.

12. Tatiana Zerjal et al., "The Genetic Legacy of the Mongols," *American Journal of Human Genetics* 72 (2003): 717–21.

13. Desmond Morris, *The Naked Ape: A Zoologist's Study of the Human Animal* (London: Jonathan Cape, 1969).

14. Nina G. Jablonski, "The Naked Truth," *Scientific American*, February 2010, pp. 42–49.

15. D. Michael Stoddart, *The Scented Ape: The Biology and Culture of Human Odour* (Cambridge: Cambridge University Press, 1990).

16. Stephen Jay Gould, *Ontogeny and Phylogeny* (Cambridge, MA: Harvard University Press, 1977). In his novel *After Many a Summer*, Aldous Huxley explored the possibility that if humans lived for a couple of hundred years, they would look like chimps.

17. One exception, according to the Old Testament, was the ancient Israelites, who believed that God wanted them to kill all the women and livestock as well. Once, when the Israelite army failed to kill the livestock, God got really angry at them.

18. Anne F. Rogers and Barbara R. Duncan, eds., *Culture, Crisis, and Conflict: Cherokee British Relations 1756–1765* (Cherokee, NC: Museum of the Cherokee Indian Press, 2009).

19. Social changes have probably made it impossible for modern society to return to these conditions. The few places in which women were kept in virtual slavery, such as Afghanistan under the Taliban, have experienced stressful changes bringing them unevenly toward modern freedom. Margaret Atwood, in *The Handmaid's Tale*, has used fiction to explore the possibility of a religiously motivated return to social conditions in which women were only a resource for men.

20. Geoffrey Miller, *The Mating Mind: How Sexual Choice Shaped the Evolution of Human Nature* (New York: Doubleday, 2000).

21. T. D. Berger and Erik Trinkaus, "Patterns of Trauma among the Neandertals," *Journal of Archaeological Science* 22 (1995): 841–52.

22. Ralph Waldo Emerson, "Nature," chap. 4, accessed February 21, 2010, http://www.emersoncentral.com/nature.htm.

23. Miller, *Mating Mind.*

24. Robin Dunbar, *Grooming, Gossip, and the Evolution of Language* (Cambridge, MA: Harvard University Press, 1998).

25. Judges 12:5–6.

26. The Acheulean technology is named after a location in France, St. Acheul, where tools of *Homo heidelbergensis* were found.

27. Richard G. Klein and Blake Edgar, *The Dawn of Human Culture* (Hoboken, NJ: Wiley, 2002), p. 106.

28. Matt Ridley, *The Agile Gene: How Nature Turns on Nurture* (New York: Harper, 2004), p. 224.

29. Darwin, *Descent of Man.*

30. Adrian Desmond and James Moore, *Darwin's Sacred Cause: How a Hatred of Slavery Shaped Darwin's Views on Human Evolution* (Boston: Houghton Mifflin, 2009).

31. Steve Olson, *Mapping Human History: Discovering the Past through Our Genes*, chap. 13 (Boston: Houghton Mifflin, 2002).

32. V. S. Johnston, *Why We Feel: The Science of Human Emotions* (Cambridge, MA: Perseus, 1999); J. T. Manning and D. Scutt, "Symmetry and Ovulation in Women," *Human Reproduction* 11 (1996): 2477–80.

33. Richard D. Alexander and Katherine M. Noonan, "Concealment of Ovulation, Parental Care, and Human Social Evolution," in *Evolutionary Biology and Human Social Behavior*, edited by Napoleon A. Chagnon and William Irons (North Scituate, MA: Duxbury, 1979), pp. 436–53.

34. David P. Barash and Judith Eve Lipton, *How Women Got Their Curves and Other Just-So Stories* (New York: Columbia University Press, 2009); David P. Barash and Nanelle R. Barash, *Madame Bovary's Ovaries: A Darwinian Look at Literature* (Concord, CA: Delta Books, 2006); Natalie Angier, *Woman: An Intimate Geography* (New York: Anchor, 1999); Jared Diamond, *Why Is Sex Fun?: The Evolution of Human Sexuality* (New York: Basic Books, 1997).

35. Sarah Blaffer Hrdy, *The Woman That Never Evolved* (Cambridge, MA: Harvard University Press, 1981); Nancy Burley, "The Evolution of Concealed Ovulation," *American Naturalist* 114 (1979): 835–58.

36. Helen Fisher, *Why We Love: The Nature and Chemistry of Romantic Love* (New York: Henry Holt, 2004); Virginia Morell, "Animal Attraction: It's His Show, but It's Her Choice," *National Geographic*, July 2003, pp. 28–55; Lauren Slater, "Love: The Chemical Reaction," *National Geographic*, February 2006, pp. 32–49.

37. A. Bartels and S. Zeki, "The Neural Basis of Romantic Love," *NeuroReport* 2 (2000): 12–15; Fisher, *Why We Love*, pp. 182–83.

CHAPTER 6

1. Michael Shermer, *The Science of Good and Evil: Why People Cheat, Gossip, Care, Share, and Follow the Golden Rule* (New York: Henry Holt, 2004).

2. Gregory S. Paul, "Cross-National Correlations of Quantifiable Societal Health with Popular Religiosity and Secularism in the Prosperous Democracies: A First Look," *Journal of Religion and Society* 5 (2007): 1–17.

3. Ara Norenzayan and Azim F. Shariff, "The Origin and Evolution of Religious Prosociality," *Science* 322 (2008): 58–62.

4. Ernst Mayr, *The Growth of Biological Thought: Diversity, Evolution, and Inheritance* (Cambridge, MA: Harvard University Press, 1982), p. 386.

5. Ralph Waldo Emerson, "The Young American," accessed February 15, 2010, http://www.emersoncentral.com/youngam.htm.

6. This Haldane quote is in wide circulation, but there appears to be no definitive source for it. It was an informal statement Haldane made in a pub. It is quoted in Lee Alan Dugatkin, "Inclusive Fitness Theory from Darwin to Hamilton," *Genetics* 176 (2007): 1375–80. However, John Maynard Smith and Eörs Szthmáry claimed that Haldane said ten cousins, not eight, just to be on the safe side, in *The Origins of Life: From the Birth of Life to the Origins of Language* (Oxford: Oxford University Press, 1999), p. 125.

7. William D. Hamilton, "Altruism and Related Phenomena, Mainly in the Social Insects," *Annual Review of Ecology and Systematics* 3 (1972): 193–232; Martin Daly and Margo Wilson, *The Truth about Cinderella: A Darwinian View of Parental Love* (New Haven, CT: Yale University Press, 1999).

8. Heather R. Mattila and Thomas D. Seeley, "Genetic Diversity in Honey Bee Colonies Enhances Productivity and Fitness," *Science* 317 (2007): 362–64.

9. Sarah Blaffer Hrdy, *Mothers and Others: The Evolutionary Origins of Mutual Understanding* (Cambridge, MA: Harvard University Press, 2009).

10. Paul W. Sherman, "Nepotism and the Evolution of Alarm Calls," *Science* 197 (1977): 1246–53.

11. Richard Ellis Hudson et al., "Altruism, Cheating, and Anti-Cheater Adaptations in Cellular Slime Molds," *American Naturalist* 160 (2002): 31–43.

12. Susan A. Dudley and Amanda L. File, "Kin Recognition in an Annual Plant," *Biology Letters* 3 (2007): 435–38.

13. Bert Hölldobler and Edward O. Wilson, *The Superorganism: The Beauty, Elegance, and Strangeness of Insect Societies* (New York: Norton, 2008).

14. Carsten Wiuf and Jotun Hein, "On the Number of Ancestors to a DNA Sequence," *Genetics* 147 (1997): 1459–68.

15. Frans De Waal, *The Age of Empathy: Nature's Lessons for a Kinder Society* (New York: Harmony Books, 2009); Nigel Barber, *Kindness in a Cruel World: The Evolution of Altruism* (Amherst, NY: Prometheus Books, 2004).

16. Dale J. Langford et al., "Social Modulation of Pain as Evidence for Empathy in Mice," *Science* 312 (2006): 1967–70.

17. J. Buhl et al., "From Disorder to Order in Marching Locusts," *Science* 312 (2006): 1402–1406.

18. Katy Payne, *Silent Thunder: In the Presence of Elephants* (New York: Penguin, 1999).

19. De Waal, *Age of Empathy*, pp. 138–39.

20. Gilbert White, *A Natural History of Selbourne*. First published in 1789.

21. Kirstin Hawkes, "Grandmothering, Menopause, and the Evolution of Human Life Histories," *Proceedings of the National Academy of Sciences USA* 95 (1998): 1–4; Sarah Blaffer Hrdy, *Mother Nature: Maternal Instincts and How They Shape the Human Species* (New York: Ballantine, 2000).

22. Natalie Angier, *Woman: An Intimate Geography* (New York: Anchor, 1999).

23. Alex Thornton and Katherine McAuliffe, "Teaching in Wild Meerkats," *Science* 313 (2006): 227–29.

24. Ursula Belugi and Marie I. St. George, *Journey from Cognition to Brain to Gene: Perspectives from Williams Syndrome* (Cambridge, MA: MIT Press, 2001).

25. Evolutionary scientists disagree on which levels—gene, individual, or group—natural selection may work. The evolution of altruism is one of the topics on which they disagree most. For a defense of the gene and individual levels, see Richard Dawkins, *The Selfish Gene* (New York: Oxford University Press, 1976) and *The Extended Phenotype: The Long Reach of the Gene* (New York: Oxford University Press, 1999). For a defense of the group level, see Elliott Sober and David Sloan Wilson, *Unto Others: The Evolution and Psychology of Unselfish Behavior* (Cambridge, MA: Harvard University Press, 1999). More recently, some evolutionary scientists have proposed that natural selection works on multiple levels: see David Sloan Wilson and

Edward O. Wilson, "Evolution 'For the Good of the Group,'" *American Scientist* 96 (2008): 380–89. David Sloan Wilson and Edward O. Wilson are not related.

26. De Waal, *Age of Empathy*, p. 187.

27. Friederike Range et al., "Effort and Reward: Inequity Aversion in Domestic Dogs?" *Proceedings of the National Academy of Sciences USA* 106 (2009): 340–45.

28. Amy Argetsinger, "The Animal Within," *Washington Post*, May 24, 2005, cited in De Waal, *Age of Empathy*.

29. Related by Sue Savage-Rumbaugh in Frans De Waal, *Bonobo: The Forgotten Ape* (Berkeley: University of California, 1997), p. 41.

30. Joan Roughgarden, *Evolution's Rainbow: Diversity, Gender, and Sexuality in Nature and People* (Berkeley: University of California Press, 2009).

31. Dave Grossman, *On Killing: The Psychological Cost of Learning to Kill in War and Society* (New York: Back Bay Books, 1995).

32. The John Updike novel *Terrorist* explores the gradual psychological entanglement of a fictitious terrorist in New Jersey.

33. William Bartram, *Travels through North and South Carolina, Georgia, East and West Florida* (London, 1792; repr., Savannah, GA: Beehive Press, 1973).

34. Michael Balter, "The Tangled Roots of Agriculture," *Science* 327 (2010): 404–406.

35. Maureen Dowd, "I Ponied Up for Sheryl Crow?," *New York Times*, February 25, 2009.

36. Joseph E. Stiglitz, "Moral Bankruptcy," *Mother Jones*, January/February 2010, pp. 28–31.

37. Kevin Drum, "Capital City," *Mother Jones*, January/February 2010, pp. 36–43.

CHAPTER 7

1. Bill Wallauer, "Do Chimps Feel Reverence for Nature?," Jane Goodall Institute of Canada, accessed February 16, 2010, http://www.janegoodall.ca/about-chimp-behaviour-waterfall.php.

2. Ian Tattersall and Jeffrey Schwartz, *Extinct Humans* (New York: Basic Books, 2001).

3. Jill Bolte Taylor, *My Stroke of Insight: A Brain Scientist's Personal Journey* (New York: Penguin, 2009).

4. Andrew Newberg, Eugene D'Aquili, and Vince Rause, *Why God Won't Go Away: Brain Science and the Biology of Belief* (New York: Ballantine, 2001).

5. Ara Norenzayan and Azim F. Shariff, "The Origin and Evolution of Religious Prosociality," *Science* 322 (2008): 58–62.

6. Sam Parnia, *What Happens When We Die?: A Groundbreaking Study into the Nature of Life and Death* (Carlsbad, CA: Hay House, 2007). Note added in press: The author has not had time to review a recent book, *Evidence of the Afterlife: The Science of the Near-Death Experience*, by Jeffrey Long, MD, and Paul Perry, published in 2010 by HarperOne.

7. There is an entire journal devoted to this subject: *Journal of Near-Death Studies*. Brain scientist Michael Persinger has been able to simulate some aspects of the near-death experience with a helmet (popularly called the God Helmet) that stimulates the brain with electromagnetic fields. Michael A. Persinger, "Religious and Mystical Experiences as Artifacts of Temporal Lobe Function: A General Hypothesis," *Perceptual and Motor Skills* 57 (1983): 1255–62; Michael A. Persinger, "Paranormal and Religious Beliefs May Be Mediated Differentially by Subcortical and Cortical Phenomenological Processes of the Temporal (Limbic) Lobes," *Perceptual and Motor Skills* 76 (1993): 247–51.

8. David Lewis-Williams, *The Mind in the Cave* (London: Thames & Hudson, 2002).

9. Leon Festinger et al., *When Prophecy Fails: A Social and Psychological Study of a Modern Group That Predicted the Destruction of the World* (New York: Harper, 1956).

10. The Heaven's Gate cult, though most of its members committed suicide, still has a website, which includes an anthology of materials titled "How and When 'Heaven's Gate' May Be Entered," accessed February 16, 2010, http://www.heavensgate.com/book/book.htm.

11. Deborah Layton, *Seductive Poison: A Jonestown Survivor's Story of Life and Death in the Peoples Temple* (New York: Anchor, 1999); Tim Reiterman and John Jacobs, *Raven: The Untold Story of the Rev. Jim Jones and His People* (New York: Dutton, 1982); Marshall Kilduff and Ron Javers, *The Suicide Cult: The Inside Story of the Peoples Temple Sect and the Massacre in Guyana* (New York: Bantam Books, 1978).

12. Richard Dawkins, "Viruses of the Mind," in *A Devil's Chaplain: Reflections on Hope, Lies, Science, and Love* (Boston: Houghton Mifflin, 2003).

13. Frank Schaeffer, *Crazy for God: How I Grew Up as One of the Elect, Helped Found the Religious Right, and Lived to Take All (or Almost All) of It Back* (Cambridge, MA: Da Capo Press, 2008). An example of one of Frank Schaeffer's political screeds from earlier decades is Franky Schaeffer, ed., *Is Capitalism Christian?* (Westchester, IL: Crossway Books, 1985).

14. James Randi, *The Faith Healers* (Amherst, NY: Prometheus Books, 1989); "Oral Roberts Tells of Talking to 900-Foot Jesus," *Tulsa World*, October 16, 1980; Justin Juozapavicius, "Oral Roberts' Son Accused of Misspending," *Washington Post*, November 9, 2007. The Duchy of Grand Fenwick is a tiny imaginary country invented by novelist Leonard Wibberley.

15. Lewis-Williams, *The Mind in the Cave*.

16. Robertson also claimed that his special protein shake allowed him, even at an advanced age, to leg-press a ton. I explored the possibilities of this statement in "Pat Robertson's Secret Ingredient," *Skeptical Inquirer* (March/April 2007): 64.

17. The earliest creed that spells out Christian theology was the Nicene Creed, approved by a church council in Nicaea (in modern Turkey) in 325 CE and recited in churches around the world today.

18. Alfred A. Cave. *The Pequot War* (Amherst: University of Massachusetts, 1996), p. 152.

19. Edward O. Wilson, *Biophilia* (Cambridge, MA: Harvard University Press, 1986).

20. Simon Conway Morris, *Life's Solution: Inevitable Humans in a Lonely Universe* (Cambridge: Cambridge University Press, 2003).

21. Frans De Waal, *The Age of Empathy: Nature's Lessons for a Kinder Society* (New York: Harmony Books, 2009).

22. However, the claim made by some writers that Jesus went to India is based on flimsy evidence. Alan Jacobs, *When Jesus Lived in India* (London: Watkins, 2009).

23. Lynn White Jr., "The Historical Roots of Our Ecologic Crisis," *Science* 155 (1967): 1203–07.

CHAPTER 8

1. Peter Pringle, *The Murder of Nikolai Vavilov: The Story of Stalin's Persecution of One of the Great Scientists of the Twentieth Century* (New York: Simon & Schuster, 2008); Gary Paul Nabhan, *Where Our Food Comes From: Retracing Nikolay Vavilov's Quest to End Famine* (Covelo, CA: Island Press, 2008); David Joravsky, *The Lysenko Affair* (Chicago: University of Chicago Press, 1986).

2. Francis Collins, *The Language of God: A Scientist Presents Evidence for Belief* (New York: Free Press, 2007).

3. In the last chapter of the second book of Samuel in the Old Testament, King David confronts, on the threshing floor of Araunah the Jebusite, the Angel of Death that has been causing an epidemic in the nation of Israel. The epidemic killed seventy thousand men, plus uncounted women and children.

4. Stanley A. Rice, "Faithful in the Little Things: Creationists and 'Operations Science,'" *Creation/Evolution* 25 (1989): 8–14.

5. Stanley A. Rice, "Scientific Creationism: Adding Imagination to Scripture," *Creation/Evolution* 24 (1988): 25–36.

6. Herbert Benson, MD, et al., "Study of the Therapeutic Effects of Intercessory Prayer (STEP) in Cardiac Bypass Patients: A Multicenter Randomized Trial of

Uncertainty and Certainty of Receiving Intercessory Prayer," *American Heart Journal* 151 (2006): 934–42.

7. C. S. Lewis, *Miracles* (New York: HarperCollins, 2001).

8. Mike Adams, "Vaccines Cause Autism: Supporting Evidence," accessed February 19, 2010, http://www.naturalnews.com/027178_vaccines_autism_Natural-Pedia.html. The scientific article on which the supposed vaccine-autism link was based has been retracted from publication. Gardiner Harris, "Journal Retracts 1998 Paper Linking Autism to Vaccines," *New York Times*, February 2, 2010.

9. Linda O. McMurry, *George Washington Carver: Scientist and Symbol* (New York: Oxford University Press, 1981).

CHAPTER 9

1. William Paddock and Paul Paddock, *Famine 1975! America's Decision: Who Will Survive?* (New York: Little, Brown, 1967).

2. Leon Hesser, *The Man Who Fed the World: Nobel Peace Prize Laureate Norman Borlaug and His Battle to End World Hunger* (Dallas: Durban House, 2006).

3. Quoted in Richard P. F. Holt et al., eds., *Post Keynesian and Ecological Economics: Confronting Environmental Issues* (Williston, VT: Edward Elgar Publishing, 2010).

4. Brandon Keim, "Vertical Farming: Apple Store Meets Greenhouse Meets Skyscraper," accessed February 21, 2010, http://www.wired.com/wiredscience/2007/12/is-the-world-re/.

5. Population Reference Bureau, accessed February 21, 2010, http://www .prb/org.

6. Garrett Hardin, "Nobody Ever Dies of Overpopulation," *Science* 171 (1971): 527.

7. Lester Brown, *The Twenty-Ninth Day* (New York: Norton, 1978).

8. Garrett Hardin, *Living within Limits: Ecology, Economics, and Population Taboos* (New York: Oxford University Press, 1995).

9. "Climate Change 2007: Synthesis Report," Intergovernmental Panel on Climate Change, accessed February 21, 2010, http://www.ipcc.ch/publications_and_data/ar4/syr/en/contents.html.

10. "National Security and the Threat of Climate Change," CNA Corporation, accessed February 21, 2010, http://securityandclimate.cna.org.

11. "Over the Top," *Science* 326 (2009): 25.

12. Stanley A. Rice, *Green Planet: How Plants Keep the Earth Alive*, chap. 3 (New Brunswick, NJ: Rutgers University Press, 2009).

13. Stefan Rahmsdorf et al., "Recent Climate Observations Compared to Projections," *Science* 316 (2007): 709.

14. Rice, *Green Planet*, chap. 12.

15. Republicans for Environmental Protection, accessed February 21, 2010, http://www.repamerica.org.

16. The opposition of Republican leadership to environmental issues is much stronger now than in the past. In 1984, President Ronald Reagan said, "If we've learned any lessons during the past few decades, perhaps the most important is that preservation of our environment is not a partisan challenge; it's common sense. Our physical health, our social happiness, and our economic well-being will be sustained only by all of us working in partnership as thoughtful, effective stewards of our natural resources." Quoted by David Jenkins, "Graham Deserves Praise, Not Censure, for Climate Change Stance," accessed February 21, 2010, http://www.repamerica.org/opinions/op-eds/132.html.

17. Andrew Grice, "Bush to G8: 'Goodbye from the World's Biggest Polluter,'" *Independent* (UK), July 10, 2008, accessed February 21, 2010, http://www.independent.co.uk/news/world/politics/bush-to-g8-goodbye-from-the-worlds-biggest-polluter-863911.html.

18. United States Climate Action Partnership, accessed February 21, 2010, http://www.us-cap.org.

19. See my essays at "A Quiet Stand of Alders," accessed February 21, 2010, http://www.stanleyrice.com.

20. James Lovelock, *The Vanishing Face of Gaia: A Final Warning* (New York: Basic Books, 2009).

21. Peter Ward and Alexis Rockman, *Future Evolution: An Illuminated History of Life to Come* (New York: Henry Holt, 2001); Stephen R. Palumbi, *The Evolution Explosion: How Humans Cause Rapid Evolutionary Change* (New York: Norton, 2001).

22. Alan Weisman, *The World without Us*, chap. 1 (New York: St. Martin's, 2007).

23. Bill McKibben, *Enough: Staying Human in an Engineered Age* (New York: St. Martin's, 2004).

24. Frank J. Tipler, *The Physics of Immortality: Modern Cosmology, God and the Resurrection of the Dead* (New York: Anchor, 1997).

25. Martin Beech, *Terraforming: The Creating of Habitable Worlds* (New York: Springer, 2009).

26. In his novel *Toward the End of Time*, John Updike describes a gigantic space station that was abandoned when war disrupted the economies of nations on Earth; the space colonists died.

27. Gene E. Likens et al., "Effects of Forest Cutting and Herbicide Treatment on Nutrient Budgets in the Hubbard Brook Watershed-Ecosystem," *Ecological Monographs* 40 (1970): 23–47.

28. Taikan Oki and Shinjiro Kanae, "Global Hydrological Cycles and World Water Resources," *Science* 313 (2006): 1068–1072.

29. Rice, *Green Planet*, p. 248.

30. Robert Costanza et al., "The Value of the World's Ecosystem Services and Natural Capital," *Nature* 387 (1997): 253–60.

31. Rice, *Green Planet*, p. 247.

BIBLIOGRAPHY

Adams, Mike. "Vaccines Cause Autism: Supporting Evidence." Accessed February 19, 2010. http://www.naturalnews.com/027178_vaccines_autism_NaturalPedia.html.

Alexander, Richard D., and Katherine M. Noonan. "Concealment of Ovulation, Parental Care, and Human Social Evolution." In *Evolutionary Biology and Human Social Behavior*, 436–53. Edited by Napoleon A. Chagnon and William Irons. North Scituate, MA: Duxbury, 1979.

Alvarez, Walter. *T. Rex and the Crater of Doom*. Princeton, NJ: Princeton University Press, 1997.

Andersson, Malte. "Female Choice Selects for Extreme Tail Length in a Widowbird." *Nature* 299 (1982): 818–20.

Angier, Natalie. *Woman: An Intimate Geography*. New York: Anchor, 1999.

Argetsinger, Amy. "The Animal Within." *Washington Post*, May 24, 2005.

Balter, Michael. "The Tangled Roots of Agriculture." *Science* 327 (2010): 404–406.

Barash, David P., and Judith Eve Lipton. *How Women Got Their Curves and Other Just-So Stories*. New York: Columbia University Press, 2009.

Barash, David P., and Nanelle R. Barash. *Madame Bovary's Ovaries: A Darwinian Look at Literature*. Concord, CA: Delta Books, 2006.

Barber, Nigel. *Kindness in a Cruel World: The Evolution of Altruism*. Amherst, NY: Prometheus Books, 2004.

Barrick, J. E., et al. "Genome Evolution and Adaptation in a Long-Term Experiment with *Escherichia coli*." *Nature* 461 (2009): 1243–47.

Bartels, A., and S. Zeki. "The Neural Basis of Romantic Love." *NeuroReport* 2 (2000): 12–15.

Bartram, William. *Travels through North and South Carolina, Georgia, East and West Florida*. London, 1792. Reprint, Savannah, GA: Beehive Press, 1973.

Bearhop, Stuart, et al. "Assortative Mating as a Mechanism for Rapid Evolution of a Migratory Divide." *Science* 310 (2005): 502–504.

Becerra, Arturo, et al. "The Very Early Stages of Biological Evolution and the Nature of the Last Common Ancestor of the Three Major Cell Domains." *Annual Review of Ecology, Evolution, and Systematics* 38 (2007): 361–79.

Beech, Martin. *Terraforming: The Creating of Habitable Worlds*. New York: Springer, 2009.

Beerling, David. *The Emerald Planet: How Plants Changed Earth's History*. Oxford: Oxford University Press, 2007.

Belugi, Ursula, and Marie I. St. George. *Journey from Cognition to Brain to Gene: Perspectives from Williams Syndrome*. Cambridge, MA: MIT Press, 2001.

Benson, Herbert, MD, et al. "Study of the Therapeutic Effects of Intercessory Prayer (STEP) in Cardiac Bypass Patients: A Multicenter Randomized Trial of Uncertainty and Certainty of Receiving Intercessory Prayer." *American Heart Journal* 151 (2006): 934–42.

Benton, Michael J. *When Life Nearly Died: The Greatest Mass Extinction of All Time*. London: Thames and Hudson, 2003.

Berger, T. D., and Erik Trinkaus. "Patterns of Trauma among Neandertals." *Journal of Archaeological Science* 22 (1995): 841–52.

Bojowald, Martin. "Follow the Bouncing Universe." *Scientific American*, October 2008, 44–51.

Brown, Lester. *The Twenty-Ninth Day*. New York: Norton, 1978.

Bryson, Bill. *A Short History of Nearly Everything*. New York: Random House, 2003.

Buhl, J., et al. "From Disorder to Order in Marching Locusts." *Science* 312 (2006): 1402–1406.

Burger, William C. *Flowers: How They Changed the World*. Amherst, NY: Prometheus Books, 2006.

Burley, Nancy. "The Evolution of Concealed Ovulation." *American Naturalist* 114 (1979): 835–58.

Calaprice, Alice. *The New Quotable Einstein*. Princeton, NJ: Princeton University Press, 2005.

Carroll, Sean. *From Eternity to Here: The Quest for the Ultimate Theory of Time*. New York: Dutton, 2010.

Carson, Rachel. *Silent Spring*. New York: Houghton Mifflin, 1962.

Catling, David C., and Kevin J. Zahnle. "The Planetary Air Leak." *Scientific American*, May 2009, 36–43.

Cave, Alfred A. *The Pequot War*. Amherst: University of Massachusetts, 1996.

Centers for Disease Control and Prevention. "Antibiotic/Antimicrobial Resistance." Accessed February 13, 2010. http://www.cdc.gov/drugresistance/index.htm.

CNA Corporation. "National Security and the Threat of Climate Change." Accessed February 21, 2010. http://securityandclimate.cna.org.

Collins, Francis. *The Language of God: A Scientist Presents Evidence for Belief*. New York: Free Press, 2007.

Conway Morris, Simon. *The Crucible of Creation: The Burgess Shale and the Rise of Animals*. New York: Oxford University Press, 1998.

———. *Life's Solution: Inevitable Humans in a Lonely Universe*. Cambridge: Cambridge University Press, 2003.

Costanza, Robert, et al. "The Value of the World's Ecosystem Services and Natural Capital." *Nature* 387 (1997): 253–60.

Daly, Martin, and Margo Wilson. *The Truth about Cinderella: A Darwinian View of Parental Love*. New Haven, CT: Yale University Press, 1999.

Darwin, Charles. *The Descent of Man, and Selection in Relation to Sex*. New York: Penguin, 2004.

———. *On the Origin of Species by Means of Natural Selection*. 150th anniversary edition. New York: Signet, 2003.

Dawkins, Richard. *The Blind Watchmaker*. New York: Oxford University Press, 1986.

———. *The Extended Phenotype: The Long Reach of the Gene*. New York: Oxford University Press, 1999.

———. *The Selfish Gene*. New York: Oxford University Press, 1976.

———. "Viruses of the Mind." In *A Devil's Chaplain: Reflections on Hope, Lies, Science, and Love*. Boston: Houghton Mifflin, 2003.

Desmond, Adrian, and James Moore. *Darwin's Sacred Cause: How a Hatred of Slavery Shaped Darwin's Views on Human Evolution*. Boston: Houghton Mifflin, 2009.

De Waal, Frans. *The Age of Empathy: Nature's Lessons for a Kinder Society*. New York: Harmony Books, 2009.

———. *Bonobo: The Forgotten Ape*. Berkeley: University of California, 1997.

Diamond, Jared. *Why Is Sex Fun?: The Evolution of Human Sexuality*. New York: Basic Books, 1997.

Dowd, Maureen. "I Ponied Up for Sheryl Crow?" *New York Times*, February 25, 2009.

Drum, Kevin. "Capital City." *Mother Jones*, January/February 2010, 36–43.

Dudley, Susan A., and Amanda L. File. "Kin Recognition in an Annual Plant." *Biology Letters* 3 (2007): 435–38.

Dugatkin, Lee Alan. "Inclusive Fitness Theory from Darwin to Hamilton." *Genetics* 176 (2007): 1375–80.

Dunbar, Robin. *Grooming, Gossip, and the Evolution of Language*. Cambridge, MA: Harvard University Press, 1998.

Dutch, Steven. "21st Century Geocentrism." Accessed February 8, 2010. http://www.uwgb.edu/dutchs/PSEUDOSC/Geocentrism.htm.

Dykhuizen, Daniel E. "Experimental Studies of Natural Selection in Bacteria." *Annual Review of Ecology and Systematics* 21 (1990): 373–98.

Eldredge, Niles. *The Pattern of Evolution*. New York: W. H. Freeman, 1999.

Emerson, Ralph Waldo. "Nature." Accessed February 21, 2010. http://www.emersoncentral.com/nature.htm.

———. "The Young American." Accessed February 15, 2010. http://www.emersoncentral.com/youngam.htm.

Erwin, Douglas H. *Extinction: How Life on Earth Nearly Ended 250 Million Years Ago*. Princeton, NJ: Princeton University Press, 2006.

Festinger, Leon, et al. *When Prophecy Fails: A Social and Psychological Study of a Modern Group That Predicted the Destruction of the World*. New York: Harper, 1956.

Fisher, Helen. *Why We Love: The Nature and Chemistry of Romantic Love*. New York: Henry Holt, 2004.

Fonstad, Mark, William Pugatch, and Brandon Vogt. "Kansas Is Flatter Than a Pancake." *Journal of Improbable Research* 9 (2003): 16–17. Available online at http://improbable.com/airchives/paperair/volume9/v9i3/kansas.html (accessed February 8, 2010).

Gilbert, Walter. "The RNA World." *Nature* 319 (1986): 618.

Gingerich, Owen. *The Book Nobody Read: Chasing the Revolutions of Nicolaus Copernicus*. New York: Walker, 2004.

Gon, Sam, III. "The *Anomalocaris* Homepage." Accessed February 13, 2010. http://www.trilobites.info/anohome.html.

Goren-Inbar, Naama, et al. "Evidence of Hominin Control of Fire at Gesher Benot Ya'aqov, Israel." *Science* 304 (2004): 725–27.

Gould, Stephen Jay. "The Cosmic Dance of Shiva." In *The Flamingo's Smile: Reflections in Natural History*. New York: Norton, 1985.

———. *Full House: The Spread of Excellence from Plato to Darwin*. New York: Harmony Books, 1996.

———. *Ontogeny and Phylogeny*. Cambridge, MA: Harvard University Press, 1977.

———. *Wonderful Life: The Burgess Shale and the Nature of History*. New York: Norton, 1989.

Graham, Linda E., Martha E. Cook, David T. Hanson, Kathleen B. Pigg, and James M. Graham. "Structural, Physiological, and Stable Carbon Isotopic Evidence That the Enigmatic Paleozoic Fossil *Prototaxites* Formed from Rolled Liverwort Mats." *American Journal of Botany* 97 (2010): 268–75.

Grant, Peter R., and B. Rosemary Grant. *How and Why Species Multiply: The Radiation of Darwin's Finches*. Princeton, NJ: Princeton University Press, 2007.

Grice, Andrew. "Bush to G8: 'Goodbye from the World's Biggest Polluter.'" *Independent* (UK), July 10, 2008. Accessed February 21, 2010. http://www.independent.co.uk/news/world/politics/bush-to-g8-goodbye-from-the-worlds-biggest-polluter-863911.html.

Grice, Elizabeth A., et al. "Topographical and Temporal Diversity of the Human Skin Microbiome." *Science* 324 (2009): 1190–92, summarized by Marx, Christopher J. "Getting in Touch with Your Friends." *Science* 324 (2009): 1150–51.

Grossman, Dave. *On Killing: The Psychological Cost of Learning to Kill in War and Society*. New York: Back Bay Books, 1995.

Hall, Marshall. "The Non-Moving Earth and Anti-Evolution Web Page of the Fair Education Foundation, Inc." Accessed February 8, 2010. http://www.fixedearth.com.

Hamilton, William D. "Altruism and Related Phenomena, Mainly in the Social Insects." *Annual Review of Ecology and Systematics* 3 (1972): 193–32.

Hardin, Garrett. *Living within Limits: Ecology, Economics, and Population Taboos*. New York: Oxford University Press, 1995.

———. "Nobody Ever Dies of Overpopulation." *Science* 171 (1971): 527.

Harris, Gardiner. "Journal Retracts 1998 Paper Linking Autism to Vaccines." *New York Times*, February 2, 2010.

Hawkes, Kirstin. "Grandmothering, Menopause, and the Evolution of Human Life Histories." *Proceedings of the National Academy of Sciences USA* 95 (1998): 1–4.

Hazen, Robert M. *Genesis: The Scientific Quest for Life's Origin*. Washington, DC: Joseph Henry Press, 2005.

Heaven's Gate. "Heaven's Gate®: How and When It May Be Entered." Accessed February 16, 2010. http://www.heavensgate.com.

Hesser, Leon. *The Man Who Fed the World: Nobel Peace Prize Laureate Norman Borlaug and His Battle to End World Hunger*. Dallas: Durban House, 2006.

Hölldobler, Bert, and Edward O. Wilson. *The Superorganism: The Beauty, Elegance, and Strangeness of Insect Societies*. New York: Norton, 2008.

Holt, Richard P. F., et al., eds. *Post Keynesian and Ecological Economics: Confronting Environmental Issues*. Williston, VT: Edward Elgar Publishing, 2010.

Hrdy, Sarah Blaffer. *Mothers and Others: The Evolutionary Origins of Mutual Understanding*. Cambridge, MA: Harvard University Press, 2009.

———. *Mother Nature: Maternal Instincts and How They Shape the Human Species*. New York: Ballantine, 2000.

———. *The Woman That Never Evolved*. Cambridge, MA: Harvard University Press, 1981.

Hu, Yaoming, et al. "Large Mesozoic Mammals Fed on Young Dinosaurs." *Nature* 433 (2005): 149–52.

Hudson, Richard Ellis, et al. "Altruism, Cheating, and Anti-Cheater Adaptations in Cellular Slime Molds." *American Naturalist* 160 (2002): 31–43.

Huey, Raymond B., and Peter D. Ward. "Hypoxia, Global Warming, and Terrestrial Late Permian Extinctions." *Science* 308 (2005): 398–401.

Hutchinson, G. Evelyn. *The Ecological Theater and the Evolutionary Play*. New Haven, CT: Yale University Press, 1965.

Intergovernmental Panel on Climate Change. "Climate Change 2007: Synthesis Report." Accessed February 21, 2010. http://www.ipcc.ch/publications_and_data/ar4/syr/en/contents.html.

Jablonski, Nina G. "The Naked Truth." *Scientific American*, February 2010, 42–49.

Jacobs, Alan. *When Jesus Lived in India*. London: Watkins, 2009.

Jenkins, David. "Graham Deserves Praise, Not Censure, for Climate Change Stance." Accessed February 21, 2010. http://www.repamerica.org/opinions/op-eds/132.html.

Jeon, Kwang W. "Integration of Bacterial Endosymbionts in Amoebae." *International Review of Cytology* 14 (1983): 29–47.

Jet Propulsion Laboratory, National Aeronautics and Space Administration. "PlanetQuest: Exoplanet Exploration." Accessed February 10, 2010. http://planetquest.jpl.nasa.gov.

Johnson, Steven. *Emergence: The Connected Lives of Ants, Brains, Cities, and Software*. New York: Scribner, 2001.

Johnston, V. S. *Why We Feel: The Science of Human Emotions*. Cambridge, MA: Perseus, 1999.

Jones, Steve. *Y: The Descent of Men*. Boston: Houghton Mifflin, 2003.

Joravsky, David. *The Lysenko Affair*. Chicago: University of Chicago Press, 1986.

Judson, Olivia. *Dr. Tatiana's Sex Advice to All Creation*. New York: Henry Holt, 2002.

Juozapavicius, Justin. "Oral Roberts' Son Accused of Misspending." *Washington Post*, November 9, 2007.

Kalas, Paul, et al. "Optical Images of an Exosolar Planet 25 Light-Years from Earth." *Science* 322 (2008): 1345–48.

Keim, Brandon. "Vertical Farming: Apple Store Meets Greenhouse Meets Skyscraper." Accessed February 21, 2010. http://www.wired.com/wiredscience/2007/12/is-the-world-re/.

Kessler, Danny, Klaus Gase, and Ian T. Baldwin. "Field Experiments with Transformed Plants Reveal the Sense of Floral Scents." *Science* 321 (2008): 1200–1202.

Kilduff, Marshall, and Ron Javers. *The Suicide Cult: The Inside Story of the Peoples Temple Sect and the Massacre in Guyana*. New York: Bantam Books, 1978.

Klein, Richard G., and Blake Edgar. *The Dawn of Human Culture*. Hoboken, NJ: Wiley, 2002.

Knoll, Andrew H. *Life on a Young Planet: The First Three Billion Years of Evolution on Earth*. Princeton, NJ: Princeton University Press, 2003.

Lack, David. *Darwin's Finches*. Cambridge: Cambridge University Press, 1947.

Langford, Dale J., et al. "Social Modulation of Pain as Evidence for Empathy in Mice." *Science* 312 (2006): 1967–70.

Layton, Deborah. *Seductive Poison: A Jonestown Survivor's Story of Life and Death in the Peoples Temple*. New York: Anchor, 1999.

Lenski, Richard. "Experimental Evolution." Accessed February 13, 2010. https://myxo.css.msu.edu/ecoli/publist.html.

Leslie, Mitch. "On the Origin of Photosynthesis." *Science* 323 (2009): 1286–87.

Levitin, Daniel J. *This Is Your Brain on Music: The Science of a Human Obsession*. New York: Penguin, 2007.

Lewis, C. S. *Miracles*. New York: HarperCollins, 2001.

Lewis-Williams, David. *The Mind in the Cave*. London: Thames & Hudson, 2002.

Li, Quanguo, et al. "Plumage Color Patterns of an Extinct Dinosaur." *Science Express*, published online February 5, 2010. Accessed February 13, 2010. http://www.sciencemag.org/cgi/rapidpdf/science.1186290v1.pdf.

Likens, Gene E., et al. "Effects of Forest Cutting and Herbicide Treatment on Nutrient Budgets in the Hubbard Brook Watershed-Ecosystem." *Ecological Monographs* 40 (1970): 23–47.

Long, John A. *The Rise of the Fishes: 500 Million Years of Evolution*. Baltimore, MD: Johns Hopkins University Press, 1996.

Lovejoy, C. Owen, et al. "The Great Divides: *Ardipithecus ramidus* Reveals the Post-crania of Our Last Common Ancestors with African Apes." *Science* 326 (2009): 100–106.

Lovelock, James E. *Gaia: A New Look at Life on Earth*. New York: Oxford University Press, 1987.

———. *The Vanishing Face of Gaia: A Final Warning*. New York: Basic Books, 2009.

Manning, J. T., and D. Scutt. "Symmetry and Ovulation in Women." *Human Reproduction* 11 (1996): 2477–80.

Margulis, Lynn. *Symbiotic Planet: A New Look at Evolution*. New York: Basic Books, 1998.

Margulis, Lynn, and Dorion Sagan. *Acquiring Genomes: A Theory of the Origins of Species*. New York: Basic Books, 2002.

———. *Mystery Dance: On the Evolution of Human Sexuality*. New York: Simon & Schuster, 1991.

Mattila, Heather R., and Thomas D. Seeley. "Genetic Diversity in Honey Bee Colonies Enhances Productivity and Fitness." *Science* 317 (2007): 362–64.

Matz, Mikhail V., et al. "Giant Deep-Sea Protist Produces Bilaterian-Like Traces." *Current Biology* 18 (2008): 1–6.

Mayama, Satoshi, et al. "Direct Imaging of Bridged Twin Protoplanetary Disks in a Young Multiple Star." *Science* 327 (2010): 306–308.

Maynard Smith, John, and Eörs Szthmáry. *The Origins of Life: From the Birth of Life to the Origins of Language*. Oxford: Oxford University Press, 1999.

Mayr, Ernst. *The Growth of Biological Thought: Diversity, Evolution, and Inheritance*. Cambridge, MA: Harvard University Press, 1982.

McKibben, Bill. *Enough: Staying Human in an Engineered Age*. New York: St. Martin's, 2004.

McMenamin, Mark A. S. "Gaia and Glaciation: Lipalian (Vendian) Environmental Crisis." In *Scientists Debate Gaia: The Next Century*, 115–28. Edited by Stephen H. Schneider et al. Cambridge, MA: MIT Press, 2004.

———. *The Garden of Ediacara: Discovering the First Complex Life*. New York: Columbia University Press, 1998.

McMenamin, Mark A. S., and Dianna L. S. McMenamin. *Hypersea*. New York: Columbia University Press, 1996.

McMurry, Linda O. *George Washington Carver: Scientist and Symbol*. New York: Oxford University Press, 1981.

Miller, Geoffrey. *The Mating Mind: How Sexual Choice Shaped the Evolution of Human Nature*. New York: Doubleday, 2000.

Montagu, Ashley. *The Natural Superiority of Women*. 5th ed. Lanham, MD: Rowman and Littlefield, 1999.

Morell, Virginia. "Animal Attraction: It's His Show, but It's Her Choice." *National Geographic*, July 2003, 28–55.

Morris, Desmond. *The Naked Ape: A Zoologist's Study of the Human Animal*. London: Jonathan Cape, 1969.

Nabhan, Gary Paul. *Where Our Food Comes From: Retracing Nikolay Vavilov's Quest to End Famine*. Covelo, CA: Island Press, 2008.

Newberg, Andrew, Eugene D'Aquili, and Vince Rause. *Why God Won't Go Away: Brain Science and the Biology of Belief*. New York: Ballantine, 2001.

Norenzayan, Ara, and Azim F. Shariff. "The Origin and Evolution of Religious Prosociality." *Science* 322 (2008): 58–62.

Norton, Andrew P., et al. "Mycophagous Mites and Foliar Pathogens: Leaf Domatia Regulate Tritrophic Interactions in Grapes." *Ecology* 81 (2000): 490–99.

Okamoto, Noriko, and Isao Inouye. "A Secondary Symbiosis in Progress?" *Science* 310 (2005): 287.

Oki, Taikan, and Shinjiro Kanae. "Global Hydrological Cycles and World Water Resources." *Science* 313 (2006): 1068–1072.

Olson, Steve. *Mapping Human History: Discovering the Past through Our Genes*. Boston: Houghton Mifflin, 2002.

"Oral Roberts Tells of Talking to 900-Foot Jesus." *Tulsa World*, October 16, 1980.

"Over the Top." *Science* 326 (2009): 25.

Paddock, William, and Paul Paddock. *Famine 1975! America's Decision: Who Will Survive?* New York: Little, Brown, 1967.

Palmer, Todd M., et al. "Breakdown of an Ant-Plant Mutualism Follows the Loss of Large Herbivores from an African Savanna." *Science* 319 (2008): 192–95. Summarized by Leslie, Mitch. "The Importance of Being Eaten." *Science* 319 (2008): 146–47.

Palumbi, Stephen R. *The Evolution Explosion: How Humans Cause Rapid Evolutionary Change*. New York: Norton, 2001.

Parnia, Sam. *What Happens When We Die?: A Groundbreaking Study into the Nature of Life and Death*. Carlsbad, CA: Hay House, 2007.

Paul, Gregory S. "Cross-National Correlations of Quantifiable Societal Health with Popular Religiosity and Secularism in the Prosperous Democracies: A First Look." *Journal of Religion and Society* 5 (2007): 1–17.

Payne, Katy. *Silent Thunder: In the Presence of Elephants*. New York: Penguin, 1999.

Persinger, Michael A. "Paranormal and Religious Beliefs May Be Mediated Differentially by Subcortical and Cortical Phenomenological Processes of the Temporal (Limbic) Lobes." *Perceptual and Motor Skills* 76 (1993): 247–51.

———. "Religious and Mystical Experiences as Artifacts of Temporal Lobe Function: A General Hypothesis." *Perceptual and Motor Skills* 57 (1983): 1255–62.

Pollan, Michael. *The Botany of Desire: A Plant's-Eye View of the World.* New York: Random House, 2001.

Population Reference Bureau. Accessed February 21, 2010. http://www.prb/org.

Pringle, Peter. *The Murder of Nikolai Vavilov: The Story of Stalin's Persecution of One of the Great Scientists of the Twentieth Century.* New York: Simon & Schuster, 2008.

Rahmsdorf, Stefan, et al. "Recent Climate Observations Compared to Projections." *Science* 316 (2007): 709.

Randi, James. *The Faith Healers.* Amherst, NY: Prometheus Books, 1989.

Range, Friederike, et al. "Effort and Reward: Inequity Aversion in Domestic Dogs?" *Proceedings of the National Academy of Sciences USA* 106 (2009): 340–45.

Reiterman, Tim, and John Jacobs. *Raven: The Untold Story of the Rev. Jim Jones and His People.* New York: Dutton, 1982.

Remy, Winfried, Thomas N. Taylor, Hagen Hass, and Hans Kerp. "Four-Hundred-Million-Year-Old Vesicular Arbuscular Mycorrhizae." *Proceedings of the National Academy of Sciences USA* 91 (1994): 11841–43.

Ren, Dong, et al. "A Probable Pollination Mode before Angiosperms: Eurasian, Long-Proboscid Scorpionflies." *Science* 326 (2009): 840–47.

Republicans for Environmental Protection. Accessed February 21, 2010. http://www.repamerica.org.

Ricardo, Alonso, and Jack W. Szostak. "Life on Earth." *Scientific American*, September 2009, 54–61.

Rice, Stanley A. *Encyclopedia of Evolution.* New York: Facts on File, 2006.

———. "Faithful in the Little Things: Creationists and 'Operations Science.'" *Creation/Evolution* 25 (1989): 8–14.

———. *Green Planet: How Plants Keep the Earth Alive.* New Brunswick, NJ: Rutgers University Press, 2009.

———. "Pat Robertson's Secret Ingredient." *Skeptical Inquirer*, March/April 2007, 64.

———. "A Quiet Stand of Alders." Accessed February 21, 2010. http://www.stanleyrice.com.

———. "Scientific Creationism: Adding Imagination to Scripture." *Creation/Evolution* 24 (1988): 25–36.

Rico-Gray, Victor, and Paolo S. Oliveira. *The Ecology and Evolution of Ant-Plant Interactions.* Chicago: University of Chicago Press, 2007.

Ridley, Matt. *The Agile Gene: How Nature Turns on Nurture.* New York: Harper, 2004.

———. *The Red Queen: Sex and the Evolution of Human Nature.* New York: Harper, 2003.

Robinson, Jennifer M. "Lignin, Land Plants, and Fungi: Biological Evolution Affecting Phanerozoic Oxygen Balance." *Geology* 15 (1990): 607–10.

Rogers, Anne F., and Barbara R. Duncan, eds. *Culture, Crisis, and Conflict: Cherokee British Relations 1756–1765.* Cherokee, NC: Museum of the Cherokee Indian Press, 2009.

Roughgarden, Joan. *Evolution's Rainbow: Diversity, Gender, and Sexuality in Nature and People.* Berkeley: University of California Press, 2009.

Schaeffer, Frank. *Crazy for God: How I Grew Up as One of the Elect, Helped Found the Religious Right, and Lived to Take All (or Almost All) of It Back.* Cambridge, MA: Da Capo Press, 2008.

Schaeffer, Franky, ed. *Is Capitalism Christian?* Westchester, IL: Crossway Books, 1985.

Schnayerson, Michael, and Mark J. Plotkin, *The Killers Within: The Deadly Rise of Drug-Resistant Bacteria.* New York: Back Bay Books, 2003.

Schneider, Stephen H., et al., eds. *Scientists Debate Gaia: The Next Century.* Cambridge, MA: MIT Press, 2004.

Sherman, Paul W. "Nepotism and the Evolution of Alarm Calls." *Science* 197 (1977): 1246–53.

Shermer, Michael. *The Science of Good and Evil: Why People Cheat, Gossip, Care, Share, and Follow the Golden Rule.* New York: Henry Holt, 2004.

Shklovsky, Iosif S., and Carl Sagan. *Intelligent Life in the Universe.* San Francisco: Holden-Day, 1966.

Shubin, Neil. *Your Inner Fish: A Journey into the 3.5-Billion-Year History of the Human Body.* New York: Pantheon, 2008.

Slater, Lauren. "Love: The Chemical Reaction." *National Geographic*, February 2006, 32–49.

Sober, Elliott, and David Sloan Wilson. *Unto Others: The Evolution and Psychology of Unselfish Behavior.* Cambridge, MA: Harvard University Press, 1999.

Stanley, Steven M. *Children of the Ice Age: How a Global Catastrophe Allowed Humans to Evolve.* New York: Harmony Books, 1996.

Stiglitz, Joseph E. "Moral Bankruptcy." *Mother Jones*, January/February 2010, 28–31.

Stoddart, D. Michael. *The Scented Ape: The Biology and Culture of Human Odour.* Cambridge: Cambridge University Press, 1990.

Tattersall, Ian, and Jeffrey Schwartz. *Extinct Humans.* New York: Basic Books, 2001.

Taylor, Jill Bolte. *My Stroke of Insight: A Brain Scientist's Personal Journey.* New York: Penguin, 2009.

Thornton, Alex, and Katherine McAuliffe. "Teaching in Wild Meerkats." *Science* 313 (2006): 227–29.

Thurber, James, and E. B. White. *Is Sex Necessary?: Or Why You Feel the Way You Do.* 75th anniversary edition. New York: Harper, 2004.

Tipler, Frank J. *The Physics of Immortality: Modern Cosmology, God and the Resurrection of the Dead.* New York: Anchor, 1997.

Toon, Owen B., Alan Robock, Richard P. Turco, Charles Bardeen, Luke Oman, and Georgiy L. Stenchikov. "Consequences of Regional-Scale Nuclear Conflicts." *Science* 315 (2007): 1224–25.

Trefil, James, Harold J. Morowitz, and Eric Smith. "The Origin of Life." *American Scientist* 97 (2009): 206–13.

Turco, Richard P., Owen B. Toon, T. P. Ackerman, J. B. Pollack, and Carl Sagan. "Nuclear Winter: Global Consequences of Multiple Nuclear Explosions." *Science* 222 (1983): 1283–92.

Tyson, Neil deGrasse, and Donald Goldsmith. *Origins: Fourteen Billion Years of Cosmic Evolution.* New York: Norton, 2004.

United States Climate Action Partnership. Accessed February 21, 2010. http://www.us-cap.org.

Van Valen, Leigh. "A New Evolutionary Law." *Evolutionary Theory* 1 (1973): 1–30.

Verma, Surenda. *The Tunguska Fireball.* London: Icon, 2004.

Volk, Tyler. "Gaia Is Life in a Wasteworld of By-Products." In *Scientists Debate Gaia: The Next Century*, 27–36. Edited by Schneider, Stephen H., et al. Cambridge, MA: MIT Press, 2004.

Walker, Gabrielle. *Snowball Earth: The Story of a Great Global Catastrophe That Spawned Life As We Know It.* New York: Crown Publishers, 2001.

Wallauer, Bill. "Do Chimps Feel Reverence for Nature?" Jane Goodall Institute of Canada. Accessed February 16, 2010. http://www.janegoodall.ca/about-chimp-behaviour-waterfall.php.

Ward, Peter. *The Medea Hypothesis: Is Life on Earth Ultimately Self-Destructive?* Princeton, NJ: Princeton University Press, 2009.

Ward, Peter, and Alexis Rockman. *Future Evolution: An Illuminated History of Life to Come.* New York: Henry Holt, 2001.

Ward, Peter, and Donald Brownlee. *Rare Earth: Why Complex Life Is Uncommon in the Universe.* New York: Copernicus, 2000.

Weiner, Jonathan. *The Beak of the Finch: A Story of Evolution in Our Time.* New York: Knopf, 1994.

Weisman, Alan. *The World without Us.* New York: St. Martin's, 2007.

Werner, Michael W., and Michael A. Jura. "Improbable Planets." *Scientific American*, June 2009, 38–45.

White, Gilbert. *A Natural History of Selbourne.* First published in 1789.

White, Lynn, Jr. "The Historical Roots of Our Ecologic Crisis." *Science* 155 (1967): 1203–1207.

White, Tim D., et al. "Macrovertebrate Paleontology and the Pliocene Habitat of *Ardipithecus ramidus.*" *Science* 326 (2009): 87–93.

Williams, Becky L., Edmund D. Brodie Jr., and Edmund D. Brodie III. "A Resistant Predator and Its Toxic Prey: Persistence of Newt Toxin Leads to Poisonous (Not Venomous) Snakes." *Journal of Chemical Ecology* 30 (2004): 1901–19.

Willson, Mary F., and Nancy Burley. *Mate Choice in Plants: Tactics, Mechanisms, and Consequences*. Princeton, NJ: Princeton University Press, 1983.

Wilson, Christopher G., and Paul W. Sherman. "Anciently Asexual Bdelloid Rotifers Escape Lethal Fungal Parasites by Drying Up and Blowing Away." *Science* 327 (2010): 574–76.

Wilson, David Sloan, and Edward O. Wilson. "Evolution 'For the Good of the Group.'" *American Scientist* 96 (2008): 380–89.

Wilson, Edward O. *Biophilia*. Cambridge, MA: Harvard University Press, 1986.

Wiuf, Carsten, and Jotun Hein. "On the Number of Ancestors to a DNA Sequence." *Genetics* 147 (1997): 1459–68.

Woese, Carl. *The Genetic Code*. New York: Harper & Row, 1968.

Wrangham, Richard. *Catching Fire: How Cooking Made Us Human*. New York: Basic Books, 2009.

Xu, J., and J. I. Gordon. "Honor Thy Symbionts." *Proceedings of the National Academy of Sciences USA* 100 (2003): 10452–59.

Young, Erick T. "Cloudy with a Chance of Stars." *Scientific American*, February 2010, 34–41.

Yuan, Xunlai, Shuhai Xiao, and Thomas N. Taylor. "Lichen-Like Symbiosis 600 Million Years Ago." *Science* 308 (2005): 1017–20.

Zerjal, Tatiana, et al. "The Genetic Legacy of the Mongols." *American Journal of Human Genetics* 72 (2003): 717–21.

Zhang, Tao, and Dun-Yan Tan. "An Examination of the Function of Male Flowers in an Andromonoecious Shrub *Capparis spinosa*." *Journal of Integrative Biology* 51 (2009): 316–24.

Zheng, Xiao-Ting, et al. "An Early Cretaceous Heterodontosaurid Dinosaur with Filamentous Integumentary Structures." *Nature* 458 (2009): 333–36.

INDEX